ENERGY SECTOR STANDARD OF THE PEOPLE'S REPUBLIC OF CHINA
中华人民共和国能源行业标准

Code for Design of Concrete Arch Dams
混凝土拱坝设计规范

NB/T 10870-2021
Replace DL/T 5346-2006

Chief Development Department: China Renewable Energy Engineering Institute
Approval Department: National Energy Administration of the People's Republic of China
Implementation Date: June 22, 2022

China Water & Power Press
中国水利水电出版社
Beijing 2022

All rights reserved. No part of this publication may be reproduced, stored in a retrieval system, or transmitted in any form or by any means—electronic, mechanical, photocopying, recording or otherwise, without prior written permission of the publisher.

图书在版编目（ＣＩＰ）数据

混凝土拱坝设计规范 : NB/T 10870—2021 = Code for Design of Concrete Arch Dams (NB/T 10870—2021) : 英文 / 国家能源局发布. -- 北京 : 中国水利水电出版社，2022.8
  ISBN 978-7-5226-0977-5

Ⅰ. ①混… Ⅱ. ①国… Ⅲ. ①混凝土坝－拱坝－设计规范－中国－英文 Ⅳ. ①TV642.4-65

中国版本图书馆CIP数据核字(2022)第163596号

ENERGY SECTOR STANDARD
OF THE PEOPLE'S REPUBLIC OF CHINA
中华人民共和国能源行业标准

Code for Design of Concrete Arch Dams
混凝土拱坝设计规范
NB/T 10870-2021
Replace DL/T 5346-2006
（英文版）

First published 2022
Issued by National Energy Administration of the People's Republic of China
国家能源局　发布
Translation organized by China Renewable Energy Engineering Institute
水电水利规划设计总院　组织翻译
Published by China Water & Power Press
中国水利水电出版社　出版发行

Tel: (+ 86 10) 68545810　68545874
haoying@mwr.gov.cn
Account name: China Water & Power Press
Account number: 0200096319000089691
Address: No.1, Yuyuantan Nanlu, Haidian District, Beijing 100038, China
Organization code: 400014639
http: //www.waterpub.com.cn

中国水利水电出版社微机排版中心
北京中献拓方科技发展有限公司
184mm×260mm　16开本　7.25印张　176千字
2022年8月第1版　2022年8月第1次印刷

Price（定价）：￥1150.00 (US $ 165.00)

Introduction

This English version is one of China's energy sector standard series in English. Its translation was organized by China Renewable Energy Engineering Institute authorized by National Energy Administration of the People's Republic of China in compliance with relevant procedures and stipulations. This English version was issued by National Energy Administration of the People's Republic of China in Announcement [2022] No. 4 dated May 13, 2022.

This version was translated from the Chinese Standard NB/T 10870-2021, *Code for Design of Concrete Arch Dams*, published by China Water & Power Press. The copyright is reserved by National Energy Administration of the People's Republic of China. In the event of any discrepancy in the implementation, the Chinese version shall prevail.

Many thanks go to the staff from the relevant standard development organizations and those who have provided generous assistance in the translation and review process.

For further improvement of the English version, any comments and suggestions are welcome and should be addressed to:

China Renewable Energy Engineering Institute
No. 2 Beixiaojie, Liupukang, Xicheng District, Beijing 100120, China
Website: www.creei.cn

Translating organizations:

POWERCHINA Chengdu Engineering Corporation Limited
China Renewable Energy Engineering Institute

Translating staff:

HE Chunhui	XUE Lijun	WANG Haibo	TANG Xuejuan
ZHENG Fugang	LI Jinyang	LAI Changjiang	OU Wenbing
LIU Bin	ZHANG Gongping	LIU Tianwei	LEI Yifan
WU Mingxin			

Review panel members:

JIN Feng	Tsinghua University
QIAO Peng	POWERCHINA Northwest Engineering Corporation Limited
LIU Xiaofen	POWERCHINA Zhongnan Engineering Corporation

	Limited
QIE Chunsheng	Senior English Translator
ZHANG Ming	Tsinghua University
YAN Wenjun	Army Academy of Armored Forces, PLA
CHENG Jing	Hohai University
QI Wen	POWERCHINA Beijing Engineering Corporation Limited
JIA Haibo	POWERCHINA Kunming Engineering Corporation Limited
ZHANG Jing	POWERCHINA Chengdu Engineering Corporation Limited
LI Shisheng	China Renewable Energy Engineering Institute

National Energy Administration of the People's Republic of China

翻译出版说明

本译本为国家能源局委托水电水利规划设计总院按照有关程序和规定，统一组织翻译的能源行业标准英文版系列译本之一。2022年5月13日，国家能源局以2022年第4号公告予以公布。

本译本是根据中国水利水电出版社出版的《混凝土拱坝设计规范》NB/T 10870—2021翻译的，著作权归国家能源局所有。在使用过程中，如出现异议，以中文版为准。

本译本在翻译和审核过程中，本标准编制单位及编制组有关成员给予了积极协助。

为不断提高本译本的质量，欢迎使用者提出意见和建议，并反馈给水电水利规划设计总院。

地址：北京市西城区六铺炕北小街2号
邮编：100120
网址：www.creei.cn

本译本翻译单位：中国电建集团成都勘测设计研究院有限公司
　　　　　　　　　水电水利规划设计总院

本译本翻译人员：贺春晖　薛利军　王海波　汤雪娟
　　　　　　　　　郑付刚　李金洋　赖长江　欧文兵
　　　　　　　　　刘　斌　张公平　刘天为　雷艺繁
　　　　　　　　　武明鑫

本译本审核人员：

金　峰　清华大学

乔　鹏　中国电建集团西北勘测设计研究院有限公司

刘小芬　中国电建集团中南勘测设计研究院有限公司

郄春生　英语高级翻译

张　明　清华大学

闫文军　中国人民解放军陆军装甲兵学院

程　井　河海大学

齐　文　中国电建集团北京勘测设计研究院有限公司

贾海波　中国电建集团昆明勘测设计研究院有限公司
张　敬　中国电建集团成都勘测设计研究院有限公司
李仕胜　水电水利规划设计总院

                国家能源局

Announcement of National Energy Administration of the People's Republic of China
[2021] No. 6

In accordance with *Standardization Law of The People's Republic of China* and *Measures for the Administration of Energy Sector Standardization*, National Energy Administration of the People's Republic of China has approved and issued 356 energy sector standards including *Code for Engineering Design of Underground Forced Permeability Enhancement for Pre-drainage of Coal Gas* (Attachment 1) and the foreign language version of 25 energy sector standards including *Technical Code for Design and Calculation of Combustion System of Fossil-fired Power Plant* (Attachment 2).

Attachments: 1. Directory of Sector Standards
2. Directory of Foreign Language Version of Sector Standards

National Energy Administration of the People's Republic of China

December 22, 2021

Attachment 1:

Directory of Sector Standards

Serial No.	Standard No.	Title	Replaced standard No.	Adopted international standard No.	Approval date	Implementation date
...						
21	NB/T 10870-2021	Code for Design of Concrete Arch Dams	DL/T 5346-2006		2021-12-22	2022-06-22
...						

Foreword

According to the requirements of Document GNKJ [2015] No. 283 issued by National Energy Administration of the People's Republic of China, "Notice on Releasing the Plan for the Development and Revision of Energy Sector Standards in 2015", and after extensive investigation and research, summarization of practical experience, consultation of relevant advanced foreign standards, and wide solicitation of opinions, the drafting group has prepared this code.

The main technical contents of this code include: general provisions, terms, arch dam layout, arch dam concrete, actions and combinations of actions, stress analysis of arch dam, sliding stability analysis of abutment, overall stability analysis, seismic design of arch dam, foundation treatment, dam detailing, temperature control and crack prevention design, water release structures and their energy dissipation and erosion control, safety monitoring design, construction requirements, initial impoundment, operation and maintenance.

The main technical contents revised are as follows:

— Revising the scope of application of this code by changing "This code is applicable to the design of concrete arch dams of Grades 1, 2 and 3 on rock foundations in large and medium-sized hydropower and water resources projects to be constructed or renovated. For arch dams higher than 200 m or with special problems, special demonstration shall be performed. This code may be a reference to the design of concrete or roller compacted concrete arch dams of Grades 4 and 5" to "This code is applicable to the design of concrete arch dams for the construction, renovation and extension of hydropower projects".

— Adding the requirements of the concrete test items.

— Adding the chapter "Overall Stability Analysis", proposing relevant regulations, analysis methods and evaluation standards for overall stability analysis.

— Adding the chapter "Seismic Design of Arch Dam", proposing supplementary requirements and evaluation standards for seismic design of an extra-high dam on the basis of current relevant seismic standards.

— Adding the single safety factor design method for arch dam stress and abutment sliding stability in non-seismic conditions.

— Adding the chapter "Water Release Structures and Their Energy Dissipation and Erosion Control", in which the content of hydraulic design in the original code is included, and the technical contents of the layout and structural design of the dam outlets and the design of the plunge pool and the end dam are added.

— Adding the chapter "Construction Requirements", putting forward the requirements of the construction sequence, timing and methods for foundation excavation, foundation treatment, concrete placement and monitoring.

— Adding the chapter "Initial Impoundment, Operation and Maintenance", putting forward the requirements of the project progress, operation, inspection, and maintenance during the initial impoundment period, early-stage operation period and operation period.

— Revising the requirements for the selection of the foundation rock mass and incorporating the selection of the foundation rock mass into the chapter "Arch Dam Layout" as a separate section.

— Revising the characteristic values of concrete strength, material partial factors, and structural factors.

— Revising the technical standards for concrete temperature control and crack prevention design, and requirements for water cooling.

— Revising the table "Classification and Selection of Monitoring Items" in the chapter "Safety Monitoring Design".

— Deleting the appendix "Dam Temperature and Temperature Stress Calculation" in the original code.

National Energy Administration of the People's Republic of China is in charge of the administration of this code. China Renewable Energy Engineering Institute has proposed this code and is responsible for its routine management. Energy Sector Standardization Technical Committee on Hydropower Investigation and Design (NEA/TC15) is responsible for the explanation of specific technical contents. Comments and suggestions in the implementation of this code should be addressed to:

China Renewable Energy Engineering Institute
No. 2 Beixiaojie, Liupukang, Xicheng District, Beijing 100120, China

Chief development organizations:

POWERCHINA Chengdu Engineering Corporation Limited
China Renewable Energy Engineering Institute

Participating development organizations:

POWERCHINA Northwest Engineering Corporation Limited

POWERCHINA Guiyang Engineering Corporation Limited

POWERCHINA Kunming Engineering Corporation Limited

Chief drafting staff:

WANG Renkun	DANG Lincai	ZHANG Jing	ZHANG Chong
ZHAO Quansheng	PANG Mingliang	ZHAO Yonggang	RAO Hongling
ZHAO Wenguang	CHEN Liping	PAN Xiaohong	CHEN Lin
YOU Xiang	WANG Haibo	CAI Dewen	PAN Yanfang
HUANG Qing	YIN Huaan	WU Mingxin	ZHAO Yan
ZHANG Gongping	XUE Lijun	XIAO Yanliang	TANG Hu
HE Chunhui	DU Xiaokai	CHEN Yongfu	XIE Min
FAN Fuping	ZHANG Xiong	LV Dayong	SHU Yong
SHAO Jingdong	BAI Xingping	LIU Xiaoqiang	LIU Xiang
ZHANG Yan	ZHU Haixia	MU Gaoxiang	CHEN Xiaopeng
FENG Yuqiang	LI Ruiqing	PENG Juwei	

Review panel members:

ZHOU Jianping	CHEN Houqun	ZHANG Chuhan	CHEN Zuyu
LI Sheng	JIN Feng	WANG Fuqiang	FANG Guangda
YAO Shuanxi	PAN Jiangyang	WANG Guojin	XU Jianrong
XIAO Feng	LONG Qihuang	WANG Yiming	LI Linian
CHEN Jiang	YANG Qiang	LI Tongchun	YANG Bo
LU Zhongmin	LI Yuanzhong	CHENG Li	ZHAO Yi
LIU Rongli	LI Shisheng		

Contents

1	**General Provisions**	1
2	**Terms**	2
3	**Arch Dam Layout**	**5**
3.1	General Requirements	5
3.2	Foundation Rock Mass and Foundation Surface	5
3.3	Arch Dam Shape	6
3.4	Appurtenant Structures	8
4	**Arch Dam Concrete**	**10**
4.1	General Requirements	10
4.2	Concrete Strength and Durability	10
4.3	Concrete Raw Materials and Testing	12
5	**Actions and Combinations of Actions**	**14**
5.1	Actions	14
5.2	Combinations of Actions	15
6	**Stress Analysis of Arch Dam**	**17**
6.1	General Requirements	17
6.2	Analysis Methods	17
6.3	Control Criteria and Other Requirements	18
7	**Sliding Stability Analysis of Abutment**	**21**
7.1	General Requirements	21
7.2	Sliding Stability Analysis and Control Criteria	21
8	**Overall Stability Analysis**	**24**
8.1	General Requirements	24
8.2	Nonlinear Finite Element Analysis	24
8.3	Geomechanical Model Test	25
9	**Seismic Design of Arch Dam**	**27**
10	**Foundation Treatment**	**28**
10.1	General Requirements	28
10.2	Foundation Excavation	29
10.3	Foundation Consolidation Grouting and Contact Grouting	29
10.4	Grout Curtain	30
10.5	Foundation Drainage	33
10.6	Weak Zone Treatment	34
11	**Dam Detailing**	**36**
11.1	Dam Crest Elevation	36
11.2	Dam Crest Arrangement	36
11.3	Transverse and Longitudinal Joints	37
11.4	Joint Grouting	39

11.5	Gallery and Access	40
11.6	Waterstops and Drainage	42
12	**Temperature Control and Crack Prevention Design**	**45**
12.1	General Requirements	45
12.2	Control Criteria for Temperature and Thermal Stress	46
12.3	Temperature Control Measures	48
13	**Water Release Structures and Their Energy Dissipation and Erosion Control**	**51**
13.1	General Requirements	51
13.2	Layout of Water Release Structures	51
13.3	Design of Water Release Structures	53
13.4	Design of Energy Dissipation and Erosion Control	55
13.5	Plunge Pool and End Dam	57
13.6	Design of Protection Against Cavitation and Abrasion	59
13.7	Protection for Flood Discharge Atomization Area	61
14	**Safety Monitoring Design**	**62**
14.1	General Requirements	62
14.2	Monitoring Items	63
14.3	Monitoring Instrument Layout	64
15	**Construction Requirements**	**69**
15.1	Foundation Surface Excavation	69
15.2	Foundation Treatment	69
15.3	Concrete Construction	70
15.4	Monitoring	71
16	**Initial Impoundment, Operation and Maintenance**	**72**
16.1	General Requirements	72
16.2	Impoundment and Early-Stage Operation	72
16.3	Operation and Maintenance	74
16.4	Analysis and Evaluation of Dam Performance	74
Appendix A	**Uplift Calculation**	**76**
Appendix B	**Single Safety Factor Design for Arch Dam Stress by Trial-Load Method Under Non-seismic Loading**	**79**
Appendix C	**Sliding Stability Analysis of Abutment**	**80**
Appendix D	**Single Safety Factor Design for Sliding Stability of Abutment Under Non-seismic Loading**	**82**
Appendix E	**Hydraulic Formulae**	**84**
Appendix F	**Classification and Selection of Monitoring Items**	**96**
Explanation of Wording in This Code		**98**
List of Quoted Standards		**99**

1 General Provisions

1.0.1 This code is formulated with a view to standardizing the design of concrete arch dams, to ensure the design quality, and to achieve operational safety and reliability, environmental friendliness, resource conservation, technological advance, and economic rationality.

1.0.2 This code is applicable to the design of concrete arch dams for the construction, renovation and extension of hydropower projects.

1.0.3 Concrete arch dams shall be classified by height into low, medium, and high dams according to the following criteria:

 1 Low dams, when the height is below 30 m.

 2 Medium dams, when the height is between 30 m and 70 m.

 3 High dams, when the height is over 70 m. A dam with a height of 200 m or more is termed an extra-high dam.

1.0.4 Concrete arch dams shall be classified by thickness-height ratio into thin, medium thick and thick arch dams according to the following criteria:

 1 Thin arch dams, when the thickness-height ratio is less than 0.20.

 2 Medium thick arch dams, when the thickness-height ratio is between 0.20 and 0.35.

 3 Thick arch dams, when the thickness-height ratio is more than 0.35.

1.0.5 The design of a roller-compacted concrete arch dam shall comply with the current sector standard NB/T 10335, *Code for Design of Roller-Compacted Concrete Arch Dams*.

1.0.6 The reasonable service life of a concrete arch dam shall be in accordance with the current sector standard NB/T 10857, *Design Code for Reasonable Service Life and Durability of Hydropower Projects*.

1.0.7 In addition to this code, the design of concrete arch dams shall comply with other current relevant standards of China.

2 Terms

2.0.1 arch dam

water retaining structure that curves upstream in plan and transmits loads by shell action into the abutments and the riverbed foundation

2.0.2 dam height

difference in elevation between the dam crest and the lowest point of the foundation surface, excluding the deep troughs, pits or holes for local treatment

2.0.3 arch dam axis

extrados of the crest arch of a dam

2.0.4 arch dam centerline

connection line between the crown point and the focal point of the central arch

2.0.5 crown cantilever

cantilever section at the arch dam centerline

2.0.6 thickness-height ratio

ratio of the base thickness of the crown cantilever to the dam height

2.0.7 arch dam shape

geometry of the arch dam determined by the horizontal arch type, vertical cantilever section shape and dimensions

2.0.8 arch shape

type of horizontal arches, generally including single-centered circle, poly-centered circle, parabola, logarithmic spiral, hyperbola, ellipse, quadratic curve, and other non-circular curves

2.0.9 central angle of arch

angle between the normal directions of the arch centerline at the intersections with left and right arch abutments

2.0.10 overhang ratio

slope of the overhang cantilever surface

2.0.11 foundation surface

face that bears the arch dam loads on the foundation

2.0.12 sectional temperature difference

temperature difference along the thickness direction of a horizontal arch

2.0.13 equivalent linear temperature difference

temperature difference between the upstream and downstream surfaces in the equivalent linear distribution along the arch thickness, converted from the actual temperature distribution by the equal area moment

2.0.14 trial-load method

method used in arch dam stress calculation, which assumes that the whole arch dam body is composed of a horizontal arch system and a vertical cantilever system, and that the load distribution between the arch and cantilever systems is determined according to the deformation compatibility at all intersection points of the arches and cantilevers

2.0.15 effective deformation modulus

deformation modulus when a heterogeneous dam foundation in a specific area is converted into a homogeneous foundation according to the principle of equivalent deformation

2.0.16 abutment

rock mass on the banks upon which the arch dam is seated, including the rock mass in foundation and that within a certain upstream and downstream range

2.0.17 abutment stability

rock mass stability of an abutment under the action of arch dam thrust, dead load of rock, uplift pressure and earthquake

2.0.18 rigid limit equilibrium method

method used in analyzing the sliding stability by the limit equilibrium principle, assuming the possible sliding rock mass as a rigid body

2.0.19 thrust block

structure between the dam body and rock foundation that transmits the arch thrust to the rock foundation

2.0.20 gravity block

structure that bears the arch thrust using its own gravity

2.0.21 abutment pad

structure between the dam and the foundation, with a width larger than the thickness of the dam body at the corresponding part, that bears and spreads the arch thrust to the rock foundation

2.0.22 aliform dam

water retaining structure constructed on the upstream side of the abutment, to lengthen the seepage path in the rock foundation, to reduce the seepage pressure, and to improve the stress distribution of the rock foundation

2.0.23 perimeter joint

contact joints between arch dam and abutment pad

2.0.24 fillet

local structure constructed in the dam toe area, to increase the thickness of the arch or cantilever end and improve the stress distribution at the dam toe

2.0.25 flood discharge outlet

general term for the crest, middle, deep, and bottom outlets in the dam body for flood discharge

2.0.26 plunge pool

structure constructed downstream of the dam, to form a sufficient water area and depth, for the energy dissipation of ski-jump jet flow and free-fall jet flow

2.0.27 weak zone

soft rock zone, fractured zone, weathered relaxed zone, etc. in the rock foundation, which does not meet the engineering requirements

2.0.28 concrete replacement

treatment measure of excavating the weak zone in the dam foundation and backfilling it with concrete

2.0.29 water-binder ratio

ratio of the mixing water to the cementitious materials of the concrete mixture by mass

2.0.30 overall stability of arch dam

stability of the arch dam-foundation system under the actions of reservoir water, dead load, seepage pressure, earthquake, etc.

3 Arch Dam Layout

3.1 General Requirements

3.1.1 An arch dam shall be built in a relatively narrow valley and on a rock foundation with suitable geological conditions.

3.1.2 The arch dam axis shall be arranged at a position where the topography is relatively favorable and the abutments are massive on both banks.

3.1.3 The layout of an arch dam shall consider the methods of water release and energy dissipation, and the layout of project structures, as well as the influence of project construction, excavation slope, and natural slope.

3.1.4 The layout of an arch dam shall be determined by comprehensive techno-economic comparison based on the natural conditions at the dam site such as topography, geology, and hydrology, as well as the purposes of the project.

3.2 Foundation Rock Mass and Foundation Surface

3.2.1 An arch dam should be built on a hard rock mass, and the middle and lower part foundations of a high dam shall be very hard rock.

3.2.2 The foundation rock mass shall be selected through comprehensive techno-economic comparisons based on the rock mass class, foundation bearing requirements, foundation treatment measures, and slope stability evaluation results. The rock mass of Class I and Class II may be directly used, the rock mass of Class III may be directly used for a medium or low dam, and the rock mass of Class III may be used for a high dam after study.

3.2.3 The foundation rock mass shall be relatively homogeneous. The influence of the geological defects in the rock mass on the arch dam stresses and sliding stability shall be studied to comprehensively determine the depth of the foundation surface and the foundation treatment method.

3.2.4 The geometry of the foundation surface shall meet the following requirements:

1. The foundation surface in elevation shall be excavated into U- or V-shapes considering the topography and geology.

2. The abutment foundation surface shape (Figure 3.2.4) should be full-radial surface, or may be non-radial surface, or other shapes after demonstration.

3. The non-radial angle α of a non-radial excavation should not exceed

30°, and the angle β between the non-radial surface and centerline of the arch dam should not be less than 10°.

4 The elevation difference between the upstream and downstream foundation surface edges at riverbed shall not be too large, and the foundation surface should be slightly inclined upstream.

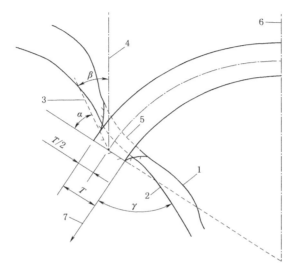

Key

1 ground line
2 contour line of the competent rock surface
3 non-radial excavation surface of the upper half of the arch abutment
4 parallel line of the arch dam centerline
5 original ground line
6 arch dam centerline
7 tangent of intrados at abutment
T thickness of arch end
$α$ non-radial angle
$β$ angle between the non-radial surface and centerline of the arch
$γ$ angle between the intrados end tangent and contour line of the competent rock

Figure 3.2.4　Schematic diagram of arch abutment

3.3　Arch Dam Shape

3.3.1 The arch dam shape design shall comprehensively consider the following factors:

1 Topographic and geological conditions at the dam site.

2 Seismic hazard and seismic requirements.

3 Layout of outlets in the dam.

4 Stress level in the dam.

5 Sliding stability of the abutment.

6 Construction materials.

7 Construction conditions.

8 Quantities of dam concrete and foundation excavation.

3.3.2 The arch dam shape design shall meet the following requirements:

1 The dam stresses, calculated using the trial-load method or the elastic finite element-equivalent stress method, shall meet the following requirements:

   1) The stress distribution is reasonable and shall meet the control criteria for compressive and tensile stresses.

   2) For a high dam with complex topographical and geological conditions, or an extra-high dam with a basic seismic intensity of Ⅵ or above, the control criteria for dam stresses shall be determined comprehensively considering the overall stability and seismic safety.

2 The horizontal arch design shall meet the following requirements:

   1) The arch shape shall be selected according to the valley shape, geological conditions, and stability conditions of the abutments. Single-centered circle, poly-centered circle, parabola, logarithmic spiral, hyperbola, ellipse, quadratic curve, other non-circular curves, or their combinations may be adopted.

   2) The maximum central angle of the arch should be 75° to 110°. A smaller angle is preferred, provided that the stress control criteria are met.

   3) The angle γ between the tangent at the intrados end and the contour line of the competent rock, as shown in Figure 3.2.4 of this code, should not be less than 30°.

   4) The arch should be thickness varied, or locally thickened at the arch end.

3 The cantilever design shall meet the following requirements:

   1) The dam face curves shall be smooth.

2) The overhang ratio shall meet the stress control criteria during operation and construction and the layout requirements of the dam discharge outlets, and should not be greater than 0.3.

3.3.3 Sensitivity analyses of significant factors, such as the effective deformation modulus of the dam foundation and temperature effects, shall be performed for a high arch dam.

3.3.4 For a sudden arch span increase near the dam crest due to the geological conditions, a gravity block, thrust block, or thrust block with an aliform dam may be set to connect the arch dam and foundations on banks.

3.3.5 For arch dams with poor stress distribution induced by the geological conditions, abutment pads, perimeter joints, or short joints at the dam heel may be set.

3.3.6 For the middle and lower toe area of a high dam, fillets may be set based on the geological conditions, the distribution of high compressive stress, and the excavation form of the dam foundation.

3.4 Appurtenant Structures

3.4.1 The arrangement of discharge outlets in the dam and their auxiliary structures shall mitigate the adverse effects on the dam stress distribution. The orifices should not be arranged in the high stress area near the dam foundation. Engineering measures shall be taken for the stress concentration zone around the orifices.

3.4.2 The energy dissipation and erosion control types and the layout of the water release structures shall not affect the stability of the dam and abutments.

3.4.3 The size, number, location and type of orifices in the dam for water intake, water supply, sediment flushing, and diversion shall be analyzed and determined according to the stress conditions of the dam and orifices, as well as the construction and operation conditions. The plugging conditions shall also be considered in the design of the diversion outlet.

3.4.4 The layout of the powerhouse at the dam toe shall mitigate the adverse effects on the structure of the arch dam, and structural joints should be set between the powerhouse and the dam.

3.4.5 The distance from the arch dam foundation surface to the underground cavern or riverbank spillway shall meet the requirement of avoiding unfavorable structural deformation or poor stress distribution in the dam.

3.4.6 The layout of the powerhouse at the dam toe, riverbank spillways, and access roads shall not affect the stability of the abutments.

3.4.7 Protective measures shall be taken for the abutments, structures, and electrical equipment affected by flood discharge atomization.

4 Arch Dam Concrete

4.1 General Requirements

4.1.1 The dam concrete shall meet the requirements for strength, deformation, thermology, permeation resistance, frost resistance, abrasion resistance, alkali-aggregate reaction resistance, and erosion resistance.

4.1.2 The quality of dam concrete shall be in accordance with the current sector standard DL/T 5144, *Specifications for Hydraulic Concrete Construction*.

4.1.3 The dam concrete strength zoning shall meet the following requirements:

   1   The concrete strength zoning shall be set according to the zoning of dam stress, outlet layout, and stress conditions of structure. The zoning of the dam stress shall be determined according to the dam stress analysis results.

   2   The types of concrete strength zones should not exceed 3.

   3   The difference in the concrete strength between adjacent zones shall not exceed 10 MPa.

   4   The portion of a high arch dam with an arch thickness less than 20 m should not be zoned along the stream direction.

4.2 Concrete Strength and Durability

4.2.1 The strength class of the dam concrete shall meet the following requirements:

   1   The strength class of the dam concrete shall be determined by the compressive strength, with a dependability of 80 %, of the 150 mm cubic specimens made and cured using the standard method, which is measured by the standard test method at the design age, notated as C_d strength, where d is the design age of the concrete. The design age of 90 d should be adopted, and the design age of 180 d or longer may also be adopted after technical demonstration.

   2   The strength class of concrete shall not be lower than $C_{90}15$.

4.2.2 The characteristic value of compressive strength of concrete may be taken from Table 4.2.2. For an extra-high arch dam, the characteristic value of dam concrete strength should be studied and checked according to the fully-graded concrete test results.

Table 4.2.2 Characteristic values of the compressive strength of dam concrete

Strength class of dam concrete	C_d15	C_d20	C_d25	C_d30	C_d35	C_d40	C_d45
Characteristic value of compressive strength (MPa)	10.0	13.4	16.7	20.1	23.4	26.8	29.6

NOTES:
1. d denotes the design age of concrete.
2. The strength class and characteristic value may be interpolated on the basis of this table.

4.2.3 The characteristic value of the tensile strength should be determined by tests, or may be taken as 0.08 to 0.10 times the compressive strength in the absence of test data.

4.2.4 The growth rates of the compressive strength at different ages should be determined by tests. In the absence of test data, the growth rates for medium or low dams may be determined according to the test data of similar projects.

4.2.5 The permeation resistance of concrete shall be determined at the design age by the test method specified in the current sector standard DL/T 5150, *Test Code for Hydraulic Concrete*, and meet the following requirements:

1. The concrete permeation resistance for a medium and low dam shall not be lower than W6.
2. The concrete permeation resistance for a high dam shall not be lower than W8.
3. The concrete permeation resistance for a high dam higher than 150 m shall not be lower than W10.

4.2.6 The frost resistance of concrete shall be determined at the design age by the test method specified in the current sector standard DL/T 5150, *Test Code for Hydraulic Concrete*. The frost resistance of concrete shall be designed according to the climate zone, freeze-thaw cycles, local microclimate conditions on the dam surface, degree of water saturation, importance of a certain part, and maintenance conditions. The frost resistance may be upgraded in the case of many adverse factors.

4.2.7 The maximum water-binder ratio of concrete shall be in accordance with Table 4.2.7.

Table 4.2.7 Maximum water-binder ratio of concrete

Climate zone	Position				
	Above water	Water level fluctuation	Underwater	Cushion	Abrasion resistance
Cold or severe cold	0.50	0.45	0.50	0.50	0.45
Mild	0.55	0.50	0.50	0.50	0.45

4.3 Concrete Raw Materials and Testing

4.3.1 The mix design of concrete shall determine the optimal mix through tests according to various property requirements, and comply with the current sector standard DL/T 5330, *Code for Mix Design of Hydraulic Concrete*.

4.3.2 The maximum particle size of dam concrete aggregates should be no less than 80 mm. Aggregates with a smaller linear expansion coefficient should be selected for the dam concrete, and the alkali-silicic acid reactive aggregate shall not be used unless demonstrated.

4.3.3 Cement selection shall meet the following requirements:

1 Moderate-heat Portland cement or Portland cement with lower heat shall be adopted for a high dam.

2 Low-heat Portland cement may be selected after special demonstration.

3 In the case of environmental water with sulfate erosion, sulfate-resistant Portland cement compliant with the current national standard GB 748, *Sulfate Resistance Portland Cement* should be selected. The selection of cementitious materials shall be specially demonstrated when other types of erosion exist in environmental water.

4 For a high dam with special demands, the chemical composition, mineral composition, hydration heat and fineness of cement may be specified, and dedicated supply cement may be determined based on productive tests.

4.3.4 Active mineral admixture should be used in dam concrete. In areas short of active mineral admixture resources, low-active or inactive mineral admixture may also be selected after test demonstration.

4.3.5 Chemical admixtures for concrete shall meet the following requirements:

1 Concrete shall be mixed with appropriate water-reducing admixture and air-entraining admixture.

2 The quality of chemical admixtures shall be in accordance with the current sector standard DL/T 5100, *Technical Code for Chemical Admixtures for Hydraulic Concrete*.

3   The chemical admixture compatibility test shall be conducted based on the selected concrete aggregates, cement, and mineral admixtures.

4.3.6   The unit weight of concrete should be determined by tests. For medium dams and low dams, or in the absence of test data, 24 kN/m³ or the values from similar projects may be taken.

4.3.7   The elastic modulus and Poisson's ratio of the high dam concrete shall be determined through tests. For medium dams and low dams, or in the absence of test data, those may be determined according to the test data of similar projects.

4.3.8   The dam concrete test items shall be selected in accordance with Table 4.3.8. The fully-graded concrete property test should be conducted for a high dam, and shall be done for an extra-high dam. The dynamic strength and dynamic elastic modulus of concrete shall be tested for a high dam of seismic design Category A.

Table 4.3.8   Dam concrete test items

Test item		Extra-high dam		High dam		Medium dam	Low dam
		Wet sieve	Fully graded	Wet sieve	Fully graded	Wet sieve	Wet sieve
Mechanical property	Compressive strength	√	√	√	*	√	√
	Tensile strength	√	√	√	*	√	√
Deformation property	Elastic modulus	√	*	√	*	√	*
	Ultimate tensile strain	√	*	√	*	√	*
	Autogenous volume change	√	*	√	*	√	*
	Shrinkage	√	*	*		*	*
	Creep	√	*	*		*	*
	Poisson's ratio	√	*	√		*	*
Thermal property	Adiabatic temperature rise	√	*	√		√	*
	Specific heat	√	*	√		*	*
	Thermal conductivity	√	*	√		*	*
	Linear expansion coefficient	√	*	√		√	*
Durability property	Frost resistance	√	√	√		√	√
	Permeation resistance	√	√	√		√	√

NOTES:
1   √ denotes mandatory.
2   * denotes optional depending on the project needs.

5 Actions and Combinations of Actions

5.1 Actions

5.1.1 The actions on an arch dam shall be selected according to the construction and operation conditions, which may include dead load, hydrostatic pressure, thermal action, uplift load, silt load, wave load, hydrodynamic load, ice load, seismic action, and other actions.

5.1.2 The calculation of the dead load shall meet the following requirements:

1 The dead load of the dam and appurtenant structures shall be determined according to the structural size and material unit weight.

2 The dead load of the appurtenant structures and permanent equipment shall be counted when the proportion in the total dead load is large.

3 The unit weight of dam concrete may be taken in accordance with Article 4.3.6 of this code.

5.1.3 The calculation of the hydrostatic pressure shall meet the following requirements:

1 The hydrostatic pressure on the upstream face of the dam shall be calculated according to the reservoir water levels under the action combinations.

2 The hydrostatic pressure on the downstream face of the dam shall be calculated according to the corresponding unfavorable downstream water level.

3 The unit weight of water should take 9.81 kN/m^3, and shall be corrected considering the sediment content for sediment-laden rivers.

5.1.4 The thermal action shall meet the following requirements:

1 The calculation of the temperature field shall comply with the current sector standard NB/T 35092, *Design Code for Temperature Control of Concrete Dam*.

2 The closure temperature field shall be comprehensively determined according to the steady temperature field, dam stress, temperature control, and crack prevention requirements.

3 The thermal action shall be determined according to the difference between the dam temperature field and the closure temperature field. The thermal action during the operation period shall meet the following requirements:

1) The normal temperature rise action shall be determined according to the difference between the closure temperature and the annual highest temperature of the dam in operation.

2) The normal temperature drop action shall be determined according to the difference between the closure temperature and the annual lowest temperature of the dam in operation.

4　The calculation of the thermal action shall comply with the current national standard GB/T 51394, *Standard for Load on Hydraulic Structures*.

5　The thermal action shall be selected in accordance with the following requirements:

1) When the trial-load method is used for arch dam stress analysis, the sectional average temperature difference and the equivalent linear temperature difference shall be considered.

2) When the finite element method is used for arch dam stress analysis, the sectional average temperature difference, equivalent linear temperature difference and non-linear temperature difference shall be considered.

5.1.5　The uplift load shall be considered in the sliding stability analysis of arch dam abutments, cracking analysis of dams and stress analysis of thick arch dams. The uplift calculation shall comply with Appendix A of this code.

5.1.6　The calculation of silt, wave, hydrodynamic, and ice loads shall comply with the current national standard GB/T 51394, *Standard for Load on Hydraulic Structures*.

5.1.7　The calculation of seismic action shall comply with the current sector standard NB 35047, *Code for Seismic Design of Hydraulic Structures of Hydropower Project*.

5.2　Combinations of Actions

5.2.1　The combination of actions shall take the most unfavorable combination of simultaneous actions for each design situation.

5.2.2　The combination of actions shall be determined from Table 5.2.2 based on the project specific conditions.

5.2.3　The calculation of action effects shall use characteristic values of actions.

Table 5.2.2 Combinations of actions

Design situation	Action combination	Design condition		Dead load	Hydrostatic load	Thermal action	Uplift load	Silt load	Wave load	Hydrodynamic load	Ice load	Seismic action
Persistent	Fundamental combination	① Normal pool level + temperature rise		√	√	√	√	√	√	–	–	–
		② Normal pool level + temperature drop		√	√	√	√	√	√	–	–	–
		③ Design flood level + temperature rise		√	√	√	√	√	√	√	–	–
		④ Dead pool level + temperature rise		√	√	√	√	√	√	–	–	–
		⑤ Dead pool level + temperature drop		√	√	√	√	√	√	–	–	–
		⑥ Freezing + temperature drop		√	√	√	√	√	–	–	√	–
Transient	Fundamental combination	★ Transverse joints partly grouted		√	–	√	–	–	–	–	–	–
		★ Transverse joints partly grouted and dam retains water		√	√	√	√	–	–	–	–	–
Accidental	Accidental combination	Check flood level + temperature rise		√	√	√	√	√	√	√	–	–
		Earthquake	Normal pool level	√	√	√	√	√	–	√	–	√
			Frequent low pool level	√	√	√	√	√	–	√	–	√

NOTE The thermal load with ★ refers to the differences between closure temperature and the temperature during the construction period.

6 Stress Analysis of Arch Dam

6.1 General Requirements

6.1.1 The stress analysis of an arch dam shall consider the influences of the following factors according to the project scale, importance, and characteristics, as well as the calculation purpose and method:

1 Layout and shape of the arch dam.

2 Complexity of the geological conditions and dam foundation treatment.

3 Dam openings and cantilever structures.

4 Stage impoundment, stage construction, construction procedures, and impounding process.

5 Closure temperature.

6 Concrete creep.

7 Embedded penstock within the dam.

8 Appurtenant structures or structural joints.

6.1.2 The following calculations shall be performed in the dam stress analysis according to specific situations, including the calculation purposes and methods, project scale, dam foundation conditions, and dam openings:

1 Principal stresses and their distribution on the upstream and downstream faces of the dam.

2 Stress distribution in typical sections.

3 Stress distribution of dam openings, water release conduits, cantilever structures, and other appurtenant structures.

4 Stresses on and beneath the dam foundation surface.

5 Stresses in construction period.

6.2 Analysis Methods

6.2.1 The trial-load method shall be used for the stress analysis of arch dam. For high dams, dams with large openings, and dams with complex foundation conditions, the elastic finite element-equivalent stress method shall be used as supplementary.

6.2.2 The elastic modulus of dam concrete shall be determined according to the test data considering the effects of concrete creep and transverse joint grouting, which may be taken as 0.6 to 0.7 times the elastic modulus of the concrete specimen.

6.2.3 The stress analysis by the trial-load method shall meet the following requirements:

1. The arch-cantilever system shall be arranged considering the engineering geological conditions. Additional arches and cantilevers should be used where the stress gradient is large.

2. For high arch dams, the degrees of freedom for deformation adjustment should be at least 4.

3. The effective deformation modulus shall be determined considering the mechanical parameters of the rock mass and the corresponding dam foundation treatment measures.

6.2.4 The computation domain and simulation objects of an arch dam using the finite element method shall meet the following requirements:

1. The computational domain shall include the dam and foundation. The extent of the dam foundation shall be able to reflect the dam-foundation interaction and include the unfavorable topographical and geological factors impacting the dam, and shall extend:

    1) Upstream of the crown cantilever not less than the dam height.

    2) Downstream of the abutments not less than 2.5 times the dam height.

    3) Left and right abutments from the foundation surface not less than 1.5 times the dam height.

    4) Downward from the foundation surface at riverbed not less than 1.5 times the dam height.

2. The modelling of the foundation should simulate the distribution of the main rock mass classes, weak zones, and main treatment measures.

3. The mesh fineness shall reflect the structure shapes and loading characteristics, and meet the accuracy requirements for calculation using the elastic finite element method.

6.3 Control Criteria and Other Requirements

6.3.1 The stress control of an arch dam under non-seismic conditions may be characterized by partial or single safety factors.

6.3.2 The arch dam stress control based on the partial factor method shall be calculated using the following formulae:

$$\gamma_0 \psi S(\cdot) \leq \frac{1}{\gamma_d} R(\cdot) \quad (6.3.2\text{-}1)$$

$$R(\cdot) = \frac{f_k}{\gamma_m} \quad (6.3.2\text{-}2)$$

where

γ_0 is the importance factor for the structure, which takes 1.10, 1.05, and 1.00 for the structures of safety Grades Ⅰ, Ⅱ, and Ⅲ, respectively;

ψ is the factor for design situation, which takes 1.00, 0.95, and 0.85 for the persistent, transient, and accidental situations, respectively;

$S(\cdot)$ is the function of action effect, namely, the principal stress calculated using the trial-load method, or equivalent stresses calculated using the elastic finite element method;

$R(\cdot)$ is the resistance function of structure;

f_k is the characteristic value of the dam concrete strength, taken as per Articles 4.2.2 and 4.2.3 of this code;

γ_m is the partial factor for the material property, taken as 1.5;

γ_d is the structural factor, taken from Table 6.3.2.

Table 6.3.2 Structural factor

Method of analysis	Stress	Fundamental combination and accidental combination
Trial-load method	Compressive	1.80
	Tensile	0.70
Elastic finite element-equivalent stress method	Compressive	1.45
	Tensile	0.55

6.3.3 The tensile stress control of the arch dam shall also meet the following requirements:

1 For persistent situations, the maximum tensile stress calculated using the trial-load method shall not exceed 1.2 MPa, and that using the elastic finite element-equivalent stress method shall not exceed 1.5 MPa.

2 For transient situations, the maximum tensile stress should not exceed 0.5 MPa before the closure grouting.

3 For accidental combination under non-seismic conditions, the maximum tensile stress should not exceed 1.5 MPa.

4 The overall stability analysis method shall be adopted to evaluate the yield range and its influence on the dam when local surface tensile stresses fail to meet the requirements.

6.3.4 Single safety factor design for the arch dam stress by the trial-load method under non-seismic loading shall comply with Appendix B of this code.

6.3.5 The stress analysis and its criteria for arch dams under seismic conditions shall comply with the current sector standard NB 35047, *Code for Seismic Design of Hydraulic Structures of Hydropower Project*.

6.3.6 The strength safety calculation for appurtenant structures of arch dams shall comply with the current sector standard DL/T 5057, *Design Specification for Hydraulic Concrete Structures*.

7 Sliding Stability Analysis of Abutment

7.1 General Requirements

7.1.1 The abutment shall meet the sliding stability requirements. Otherwise, engineering measures shall be taken to ensure the sliding stability.

7.1.2 The rigid limit equilibrium method shall be adopted for the sliding stability analysis of abutments, and the finite element method should be used as supplementary for high dams or dams with complex geological conditions.

7.1.3 The sliding stability analysis of abutments shall meet the following requirements:

1. The sliding planes shall be determined by studies on the spatial distributions, attitude, undulations and roughness, development degrees, connectivity rates, fissure fillings, and properties of the weak zones and main discontinuities around abutments.

2. The design parameters shall be determined by studies on the physical and mechanical properties of the rock mass, discontinuities, and fissure fillings in the discontinuities.

3. The uplift pressure at abutments shall be determined by studies on the hydrogeological conditions at the dam site and the seepage characteristics.

7.1.4 The sliding stability analysis of abutments shall be performed considering factors such as the arch dam layout, arch dam shape, arch end structures, water release types, foundation treatment, and construction method.

7.2 Sliding Stability Analysis and Control Criteria

7.2.1 For the sliding stability analysis of abutments, all possible sliding blocks at different elevations and locations shall be analyzed according to the spatial distribution of the block boundaries. The boundaries of the sliding blocks shall be determined in accordance with Section C.1 of this code. The sliding blocks shall be determined according to Section C.2 of this code.

7.2.2 Mechanical parameters of sliding planes and rock mass shall meet the following requirements:

1. The mechanical parameters, such as residual or peak shear strength of rock mass and discontinuities and deformation modulus of rock mass, shall be determined by tests and engineering experience. In situ tests shall be conducted for high arch dams or arch dams with complex

geological conditions.

2 The residual or peak shear strength of sliding planes shall be determined based on the test results as per Section C.3 of this code.

7.2.3 In the sliding stability analysis of abutments, the arch thrust, dead load of rock mass, uplift pressure and seismic action shall be considered. The action combinations shall comply with Article 5.2.2 of this code. The actions on sliding block shall be calculated as per Section C.4 of this code.

7.2.4 When the rigid limit equilibrium method is used in the analysis of the abutment sliding stability under non-seismic conditions, the partial factor method or single safety factor method may be adopted.

7.2.5 When the partial factor method is used for the analysis of the abutment sliding stability under non-seismic conditions, the following requirements shall be met:

1 For arch dams of Grades 1 and 2, and high arch dams, Formula (7.2.5-1) shall be satisfied.

2 For arch dams of Grade 3 and below, Formula (7.2.5-1) or Formula (7.2.5-2) shall be satisfied.

$$\gamma_0 \psi \sum T \leq \frac{1}{\gamma_{d1}} \left(\frac{\sum f'N}{\gamma'_{mf}} + \frac{\sum c'A}{\gamma'_{mc}} \right) \quad (7.2.5\text{-}1)$$

$$\gamma_0 \psi \sum T \leq \frac{1}{\gamma_{d2}} \frac{\sum fN}{\gamma_{mf}} \quad (7.2.5\text{-}2)$$

where

γ_0 is the importance factor for the structure, which takes 1.10, 1.05 and 1.00 for structures of safety Grades I, II, and III, respectively;

ψ is the factor for design situation, which takes 1.00, 0.95, and 0.85 for persistent, transient, and accidental situations, respectively;

T is the sliding force in the sliding direction (10^3 kN);

f' is the shear-friction coefficient, which shall take the average value of the peak strengths of material;

f is the friction coefficient, which takes proportion limit strength for brittle failure materials, yield strength for plastic or brittle-plastic failure materials, and residual strength for shear broken materials, respectively;

c' is the cohesion (MPa), which shall take the average value of the peak strength of material;

N is the normal force perpendicular to the sliding plane (10^3 kN);

A is the sliding plane area (m²);

γ_{d1}, γ_{d2} are the structural factors for two formulae, respectively, taken from Table 7.2.5;

$\gamma'_{mf}, \gamma'_{mc}, \gamma_{mf}$ are the partial factors for material property of the two formulae, taken from Table 7.2.5.

Table 7.2.5 Partial factors for abutment sliding stability

For Formula (7.2.5-1)	γ'_{mf}	2.40
	γ'_{mc}	3.00
	γ_{d1}	1.15
For Formula (7.2.5-2)	γ_{mf}	1.20
	γ_{d2}	1.05

7.2.6 Single safety factor design for the sliding stability of the abutment under non-seismic loading shall comply with Appendix D of this code.

7.2.7 The abutment sliding stability analysis under seismic conditions shall comply with the current sector standard NB 35047, *Code for Seismic Design of Hydraulic Structures of Hydropower Project*.

7.2.8 The abutment sliding stability shall be considered as a spatial problem, and it may be simplified to a plane problem for an arch dam below Grade 3 without specific structural planes.

7.2.9 The shallow sliding stability in foundation shall be checked for arch dams with small curvature radii, flat bank slopes, or bank slopes with joints near-parallel to the slope surface.

8 Overall Stability Analysis

8.1 General Requirements

8.1.1 The overall stability analyses shall be performed for high arch dams of Grade 1 and Grade 2 or arch dams with complex topographic and geologic conditions.

8.1.2 The nonlinear finite element method shall be adopted in the overall stability analysis. Overall geomechanical model tests should be performed for extra-high dams and high dams with complex geological conditions. Other overall stability analysis methods may also be used as supplement according to the analysis purpose.

8.1.3 The overall stability analysis of an arch dam shall cover the following:

1 Rationality of dam shape, foundation treatment, outlet layout, and structural design.

2 Overload capacity of an arch dam.

8.1.4 The conditions for the overall stability analysis of an arch dam shall include the fundamental action combination and water weight overload action combination.

8.1.5 The overall safety of an arch dam shall be comprehensively evaluated according to the overall stability analysis results and engineering analogies.

8.2 Nonlinear Finite Element Analysis

8.2.1 The computation domain and simulation objects shall meet the following requirements in addition to Item 1 of Article 6.2.4 of this code:

1 For the dam foundation, the distribution of the main rock mass classes, weak zones, and main foundation treatment measures shall be simulated, and main fissures should be simulated.

2 For the dam, the integral dam structure shall be simulated; orifices, transverse joints and other structural joints, and fillets should also be simulated.

3 The mesh fineness shall be capable of presenting the gradient of the stress distribution of structures, and the element should be of the hexahedron type. The number of element layers along the thickness direction of the dam shall meet the requirement of the analysis of the yield depth at dam heel.

8.2.2 The constitutive relationship in nonlinear finite element analysis shall suit the computation purpose and reflect the main properties of materials. The mechanical parameters of the constitutive relationship should be determined by tests, or may be determined with reference to similar projects if test conditions are not available.

8.2.3 The overall stability analysis using the nonlinear finite element method shall analyze all or part of the following items according to the specific computation purpose, computation method, project scale, dam foundation conditions, and orifice structures:

1. The range and degree of the stress concentration around the dam.
2. The range and depth of the yield of the dam.
3. The stress distribution, yield range and yield degree of the dam foundation.
4. The overall stability of controlling sliding block at the abutment.
5. The interactions of the arch dam structures with thrust blocks, gravity blocks, concrete pads, and foundation replacements.
6. The impacts of the transverse joints, perimeter joints and other specific structural joints.
7. Distribution and magnitude of deformation in the dam and foundation.

8.2.4 The fundamental combination shall be in accordance with Table 5.2.2 of this code, and should include the effects of the foundation seepage pressures, foundation weight, and reservoir and valley deformations.

8.2.5 The stresses, deformations, point safety and yield zone of the arch dam, foundation and special geological zone, as well as the rationality, shall be presented in the fundamental combination calculation.

8.2.6 In the overload case, the development of characteristic stresses and deformation in the dam and foundation, as well as the initiation, extension and connection of the yield zones in the dam heel, dam body and foundation shall be revealed based on the condition of normal pool level of fundamental combination through step-by-step overload of the water weight, and the overall safety shall be evaluated. The yield zone in dam heel shall not go beyond the curtain under 1.5 times the water weight overloading.

8.3 Geomechanical Model Test

8.3.1 The simulation objects in the geomechanical model test shall include the dam, foundation rock types, main weak zones, main fissures, and main engineering treatment measures.

8.3.2 The geomechanical model test shall include the normal load test and overload test.

8.3.3 The normal load test shall analyze the rationality of the stress, deformation and crack development to verify the design.

8.3.4 Overload tests shall study the initiation and propagation of cracks in dam and foundation and the overload capability, based on the normal loads through step-by-step overloading of water weight, to compare with the nonlinear finite element analysis results and evaluate the dam safety.

9 Seismic Design of Arch Dam

9.0.1 The seismic design of an arch dam shall comply with the current sector standard NB 35047, *Code for Seismic Design of Hydraulic Structures of Hydropower Project*.

9.0.2 An extra-high arch dam located in a region with a basic seismic intensity of VI or above shall be classified as seismic design Category A.

9.0.3 The seismic design of an extra-high arch dam shall meet the following requirements:

1. The dynamic analysis of the dam shall comply with the current sector standard NB 35047, *Code for Seismic Design of Hydraulic Structures of Hydropower Project*, and the effect of dam material nonlinearity shall also be considered.

2. Under the design earthquake, the yield zone in the dam heel shall not exceed the curtain. When local yield occurs in the other parts, the dam shall maintain normal operation after repair.

3. Under the maximum credible earthquake, the yield zone shall not go through the dam and the transverse joint shall not open to damage the waterstops, and possible residual deformations in the dam and abutments and their impacts shall be analyzed, to study the possibility of the uncontrollable release of reservoir water.

4. The seismic design shall perform the limit overload analysis of dam-foundation system, and conduct engineering analogies. This analysis should use the overload of the peak ground acceleration of design earthquake.

10 Foundation Treatment

10.1 General Requirements

10.1.1 The arch dam foundation shall be treated to meet the following requirements:

1. Sufficient strength and stiffness for load bearing and deformation control.
2. Sliding stability of abutments and overall stability of dam-foundation system.
3. Impermeability and seepage stability.
4. Durability under long-term water action.

10.1.2 The foundation treatment scheme shall be determined through structural and stability analyses according to the dam site's geological conditions and geomechanical properties of the foundation rock mass, considering the layout of adjacent structures and construction technology, and following the principle of safety, economy and effectiveness.

10.1.3 Foundation treatment measures shall be selected, such as the excavation, abutment pad, thrust block, gravity block, aliform dam, consolidation grouting, contact grouting, seepage control and drainage, concrete replacement of the weak zone, and pre-stressed anchorage, according to the actual geological conditions and treatment purposes.

10.1.4 Quality control criteria and testing methods shall be specified in the foundation treatment design.

10.1.5 The stability of the abutment slopes and the slopes above the crest shall be analyzed in accordance with the current sector standard NB/T 10512, *Code for Slope Design of Hydropower Projects*, and measures shall be taken when the stability requirements are not met.

10.1.6 The foundation treatment scheme shall be reviewed according to the conditions revealed by excavation, and may be adjusted according to specific situations.

10.1.7 The foundation treatment design in karst area shall be determined by analyses according to the ascertained karst distribution and development, the scale, distribution, connectivity, and filling of the main karst caves and conduits, as well as the water quality and activity of groundwater.

10.1.8 The uplift pressure considering the effects of the grout curtain and drainage shall be controlled within the permissible range.

10.1.9 The seepage control and drainage design of a high arch dam with complex hydrogeological conditions shall be determined considering seepage analysis.

10.2 Foundation Excavation

10.2.1 The evenness of the foundation surface of an arch dam shall be controlled according to rock mass, slope shape and construction conditions.

10.2.2 Precautionary measures and construction control measures shall be studied for relaxed rocks by foundation excavation, and engineering measures of reserving the protective layer and pre-anchorage, and construction techniques for reducing relaxation, may be adopted.

10.2.3 Geological defects exposed on the foundation surface that cannot meet the quality requirements for foundation rocks shall be removed.

10.2.4 When the replacement of weak zone extends downstream of the foundation surface to meet the requirements of load transmission, the excavation design of downstream abutment slopes shall consider the excavation and backfilling of the replacement zone.

10.2.5 The foundation rock mass prone to disintegration, weathering and softening after excavation shall be protected using reserved protective layers and concrete covers.

10.3 Foundation Consolidation Grouting and Contact Grouting

10.3.1 The foundation consolidation grouting design shall meet the following requirements:

1 The range of consolidation grouting shall be determined comprehensively according to foundation geological conditions, blasting relaxation and foundation stress, and should be properly extended upstream and downstream of the foundation surface.

2 Depths of consolidation grouting holes shall be determined according to geological conditions and stresses in the dam foundation.

3 Parameters and techniques of foundation consolidation grouting should be determined according to grouting tests and geological conditions, or may be determined by engineering analogies in the absence of grouting test data. The following requirements shall also be met:

   1) For extra-high dams, and Grade 1 and Grade 2 arch dams with complex geological conditions, the parameters and techniques shall

be determined by grouting tests.

2) The parameters and techniques shall be adjusted during construction according to construction data.

4 Foundation consolidation grouting holes should be densified and deepened near the curtain.

5 Consolidation grouting should be strengthened around the concrete replacement of weak zones.

10.3.2 Foundation consolidation grouting should be conducted in the presence of concrete cover. Otherwise, the grouting without concrete cover shall be demonstrated and appropriate measures shall be taken to ensure the quality of top rock grouting.

10.3.3 Quality inspection of consolidation grouting should adopt the elastic wave test.

10.3.4 The contact grouting design shall meet the following requirements:

1 Contact grouting shall be provided for the following parts:

1) Foundation surface with a slope greater than 50°.

2) Side walls and tops of the backfilling concrete in trenches, pits or caverns excavated in the rock foundation.

2 Contact grouting at the foundation surface should be performed combining consolidation grouting with concrete cover and curtain grouting.

10.3.5 The quality of contact grouting should be inspected using water pressure test in borehole, and may also be comprehensively evaluated using the drill core data of inspection holes, borehole TV, and grout take.

10.4 Grout Curtain

10.4.1 The grout curtain shall meet the following requirements:

1 Seepage failures at weak zones, closely jointed and fissured or poorly impermeable rocks are not allowed.

2 The grout curtain shall be continuous and durable.

3 Parameters and techniques of curtain grouting should be determined by tests considering the geological conditions, or may be determined by engineering analogies in the absence of grouting test data. The following requirements shall also be met:

1) For extra-high dams and Grade 1 and Grade 2 arch dams with complex geological conditions, the parameters and techniques shall be determined by grouting tests.

   2) The parameters and techniques shall be adjusted during construction according to construction data.

10.4.2 The seepage control criteria for a grout curtain and the permeability of the aquiclude shall meet the following requirements depending on the dam height:

1 The permeability shall not exceed 1 Lu for dams over 200 m.

2 The permeability should be 1 Lu to 3 Lu for dams between 100 m and 200 m.

3 The permeability should be 3 Lu to 5 Lu for dams between 50 m and 100 m.

4 The permeability shall not exceed 5 Lu for dams lower than 50 m.

5 For reservoirs short of water resources, the permeability should be strictly controlled.

6 In the case of poor stability conditions of abutments and slopes, the permeability should be strictly controlled.

10.4.3 The layout of a grout curtain shall meet the following requirements:

1 The curtain shall be arranged in the compression zone of the dam foundation, and its centerline may be set at a distance of 0.05 to 0.10 times the working head to the upstream dam face.

2 The curtain at abutments should extend into the aquiclude in the mountain in the short-distance direction, and shall connect the curtains of adjacent structures and those at riverbed.

3 The length of the grout curtain in abutments shall be determined comprehensively according to engineering geological, hydrogeological and topographical conditions, as well as the requirements for foundation stability and seepage control. The grout curtain should extend with a proper margin into the aquiclude or to the intersection of the normal pool level and groundwater table.

4 The grout curtain in the karst area shall be economically and reasonably aligned according to the distribution and development of karst. The axis of the grout curtain should be placed at the less-developed karst area or groundwater divide. Concrete backfilling, high pressure jet

grouting, and cutoff walls shall be adopted to block the caverns along the alignment when it is inevitably to cross karst underground rivers or conduits.

5 When there is an aquiclude shallow buried under the foundation surface, the grout curtain shall extend no less than 5 m into the aquiclude.

6 When the aquiclude under the foundation surface is deeply buried or irregularly distributed, the depth of a hanging grout curtain shall be comprehensively determined according to the project scale, geological conditions, requirements of uplift pressure control at dam base, drainage conditions, and seepage computation results, and may range from 0.3 to 0.7 times the working head upstream of the dam.

7 The hole depth of the secondary curtain in the dam foundation on the riverbed may take 1/2 to 2/3 depth of the primary curtain.

8 The upper curtain shall overlap with the lower curtain to effect closure of the curtain.

10.4.4 The number of rows, row spacing, hole spacing, and hole orientation shall be determined according to the engineering geological conditions, hydrogeological conditions, and working head, and shall meet the following requirements:

1 For sound and impermeable foundations, 1 row curtain may be adopted for dams below 100 m, and 1 to 2 rows for dams above 100 m. For poor and permeable foundations, 1 row may be adopted for dams below 50 m, and 1 to 2 rows should be adopted for dams of 50 m to 100 m, and 2 to 3 rows for dams above 100 m.

2 The curtain hole spacing should be 1.5 m to 3.0 m, and row spacing shall be slightly less than the hole spacing, and the hole spacing and row spacing should be reduced at weak zones.

3 The curtain hole should dip upstream at an angle of 75° to 90°, and penetrate the main bedding surfaces and fissures in the foundation rock mass.

10.4.5 Pure cement slurry should be adopted as the grouting material for the grout curtain, and use of fine cement slurry or chemical grouting material shall be studied when pure cement slurry cannot meet the seepage control requirements.

10.4.6 The curtain grouting pressure shall be comprehensively determined according to engineering geology, hydrogeology, working head, hole spacing, and row spacing, and shall meet the following requirements:

1. The collar grouting pressure of the hole should be 1.0 to 1.5 times the working head upstream of the dam, but no greater than 3.0 MPa.

2. The bottom grouting pressure of the hole shall not be lower than 2 times the working head upstream of the dam, but should not be greater than 6.0 MPa.

3. The grouting pressure shall gradually increase from the collar to the bottom of holes.

4. The grouting pressure shall not induce the uplift of the dam concrete and foundation rocks.

5. The grouting pressure shall not adversely affect other structures.

10.4.7 Grouting tunnels shall be arranged at different elevations when the curtain in the abutments is very deep. The layout shall meet the following requirements:

1. The layout of the grouting tunnels in abutments shall meet the requirements of curtain construction and grouting quality control.

2. The elevation difference between grouting tunnels should be 40 m to 70 m.

3. The slope of drainage ditch in grouting tunnel shall meet the requirements of free draining and waste slurry discharge.

4. The dimensions of grouting tunnels shall meet the requirements of hole layout and work space of grouting equipment, and should be 2.5 m to 4.0 m in width and 3.0 m to 4.5 m in height.

5. Grouting tunnels except those at dam crest elevation shall be connected to galleries in the dam.

10.5 Foundation Drainage

10.5.1 The layout of dam foundation drainage shall meet the following requirements:

1. One row primary drainage shall be set downstream of the grout curtain.

2. Foundation of a thick arch dam or high dam may be set with secondary drainages.

3. Abutment drainages should be set based on the stability conditions of abutments.

4   The drainage layout may be simplified for a thin arch dam of medium or low height.

10.5.2 The hole spacing should be 2 m to 3 m for primary and secondary drainages and 3 m to 6 m for abutment drainage.

10.5.3 The depth of drainage holes shall be determined according to the depths of curtain grouting and consolidation grouting, and the engineering geological and hydro-geological conditions, and shall meet the following requirements:

1   The hole depth of primary drainage should be 0.4 to 0.6 times that of the grout curtain, and shall be no less than 10 m for a medium or high dam.

2   The hole depth of secondary drainage should be 0.7 times that of primary drainage.

3   The hole depth of drainage shall be analyzed and determined when a confined aquifer or deep pervious stratum exists in dam foundation.

4   When gentle-dip discontinuities that affect the sliding stability exist in the dam foundation, the depth of drainage holes should be analyzed and determined according to their distribution.

10.5.4 The distance between primary drainage holes and curtain grouting holes shall meet the requirements of seepage stability of foundation rock, and should be no less than 2 m.

10.5.5 The invert slope of drainage tunnels shall meet the requirements of free drainage.

10.5.6 Measures shall be taken for drainage holes going through weak zones and those with unstable hole walls, to prevent local seepage failure.

10.6 Weak Zone Treatment

10.6.1 Weak zone treatment shall meet the following requirements:

1   Weak zone treatment scheme shall be comprehensively determined according to the weak zone location, attitude, width, material composition, physical and mechanical properties, and influence on dam stresses, dam foundation deformation and stability, sliding stability, and seepage stability, as well as the construction conditions.

2   Weak zone treatment scheme that significantly affects dam stresses and deformations shall be comprehensively determined through techno-economic comparisons on the basis of calculations or model tests.

3 Treatment scheme for a medium or low dam may be determined through techno-economic comparison according to engineering experience and construction conditions.

10.6.2 Weak zone treatment measures shall be determined based on the characteristics, scale and location of the weak zone and the treatment targets. The treatment measures, such as cement consolidation grouting, chemical grouting, concrete plug, concrete replacement grid by adits and shafts, shearing resistance or load transmitting structures, anchorage, or their combination, may be adopted.

10.6.3 The following measures shall be taken for weak zones with steep dips:

1 For weak zones that slightly affect the thrust transmitting, stability and deformation of foundation, the surface fractured zone may be excavated and replaced with concrete plug, and the lateral and deep rock mass may be treated with consolidation grouting.

2 For weak zones that significantly affect the strength, stability and deformation of foundation, a certain range of the dam foundation shall be treated with concrete replacement and high pressure cement grouting, or chemical grouting should be conducted if the treatment effect is unsatisfactory.

10.6.4 The treatment measures for a gentle-dip weak zone shall be determined considering the position and characteristics of the weak zone and its impact on the sliding stability, dam stress, and dam foundation deformation. Such measures as concrete replacement, shear (slide) resistant structures, anchorage, and seepage control and drainage facilities in local areas may be adopted.

10.6.5 Weak zone replacement treatment shall not damage the intact rock mass, and concrete lining or backfill should be adopted to prevent excavation-induced softening, unloading, and relaxation. The backfill, contact and consolidation grouting between concrete and foundation rock shall be enhanced. Temperature control design shall be conducted for large scale replacement concrete.

10.6.6 The seepage control treatment for a weak zone hydraulically connected to the reservoir shall comply with Article 10.4.1 of this code, and the treatment measures shall be determined according to the scale, attitude, property and working head of the weak zone. Cement, fine cement, and chemical grouting, as well as impervious wall and shaft, may be adopted.

11 Dam Detailing

11.1 Dam Crest Elevation

11.1.1 The dam crest elevation shall be higher than the highest still water level of the reservoir.

11.1.2 The elevation of the dam crest or the top of upstream parapet wall at the dam crest shall be calculated by adding the corresponding Δh to the normal pool level, design flood level, or check flood level, and the maximum elevation shall be taken. Δh shall be determined by the following formula:

$$\Delta h = h_{1\%} + h_z + h_c \tag{11.1.2}$$

where

Δh is the elevation difference between the dam crest or the parapet wall top and the still water level of the reservoir such as the normal pool level, design flood level, or check flood level (m);

$h_{1\%}$ is the wave height with a cumulative frequency of 1 %, calculated according to the current national standard GB/T 51394, *Standard for Load on Hydraulic Structures* (m);

h_z is the height from the wave centerline to the corresponding still water level, calculated according to the current national standard GB/T 51394, *Standard for Load on Hydraulic Structures* (m);

h_c is the safety margin in height, taking value according to the current sector standard DL 5180, *Classification & Design Safety Standard of Hydropower Projects* (m).

11.2 Dam Crest Arrangement

11.2.1 The parapet wall should be a reinforced concrete structure integrated with the dam body, and the wall structure shall meet the requirements of withstanding the impacts of waves and floating debris. Expansion joints and waterstops shall be set in the parapet wall at the transverse joints of the dam, and the waterstops shall be connected to those in the transverse joints of the dam.

11.2.2 The width of the dam crest shall meet the requirements of operation and transportation, and should be no less than 3 m. Corbel structures may be set on the crest of non-overflow section when necessary.

11.2.3 The service bridge, access bridge, hoist chamber, and bent frame at the overflow section shall meet the following requirements:

1 The layout of service bridge, access bridge, hoist chamber, and bent frame shall meet the requirements for the arrangement, operation and maintenance, transportation, and monitoring of gates and hoists.

2 The structures of service bridge, access bridge, hoist chamber, and bent frame shall meet the strength and stiffness requirements under the operating conditions. In regions of high seismic intensity, lightweight measures shall be adopted.

3 The service bridge and access bridge should be of a prefabricated or pre-stressed reinforced concrete structure. Sufficient clearance under the bridge shall be reserved for flood discharging.

11.2.4 The dam crest pavement shall be provided with transverse slope and drainage. The sidewalk should be 20 cm to 30 cm higher than the pavement. The downstream side of the dam crest shall be set with a guardrail which shall meet the safety requirements.

11.2.5 The cable trenches on dam crest shall meet the following requirements:

1 The layout of the cable trenches shall comprehensively consider the access of the power, monitoring, and lighting cables.

2 The cross-section and dimensions of a cable trench shall be determined according to the cable type, quantity, installation, and maintenance requirements.

3 The cable trench cover shall be capable of accommodating the vehicle load, whose structure type and block length shall meet the installation and maintenance requirements.

4 Angle steel or other protective measures should be provided for the corners of cable trenches and covers.

5 Drainage shall be set at the bottom of the cable trenches.

11.2.6 Lighting facilities on the dam crest shall meet the requirements of patrol inspection, transportation and safety.

11.2.7 The dam crest and appurtenant structures and facilities shall coordinate with the overall planning of the project structures, and shall consider the landscaping.

11.3 Transverse and Longitudinal Joints

11.3.1 The arrangement of transverse joints for the arch dam shall meet the requirements for concrete temperature control and crack prevention, along with construction intensity. The locations of transverse joints shall be comprehensively determined, considering the locations of the dam outlets and openings.

11.3.2 For a thick arch dam, longitudinal joints may be set after the demonstration considering temperature control and crack prevention during the construction period.

11.3.3 The arrangement of transverse joints shall meet the following requirements:

1. The spacing of transverse joints along the upstream arc length should be 15 m to 25 m.
2. Transverse joints should be arranged radially in the plan.
3. Transverse joint surfaces may be vertical plane or torse. The included angle between the transverse joint surface and the dam foundation surface should be no less than 60°.
4. Transverse joints shall be provided with the keys and grouting system.
5. Keys may be spherical, trapezoidal or arc, and the trapezoidal or arc key shall be vertical.

11.3.4 The arrangement of longitudinal joints shall meet the following requirements:

1. The longitudinal joint surface should be vertical and the longitudinal joints of adjacent dam blocks shall be staggered.
2. For the longitudinal joints exposed on the downstream dam face, the included angle with the dam face should be no less than 60°.
3. Circular holes and reinforcement may be set at the ends of longitudinal joints.
4. Longitudinal joints shall be provided with horizontal keys and grouting system, and the key shape should be a triangle.

11.3.5 The grouting system for the transverse and longitudinal joints shall meet the following requirements:

1. The joint shall be partitioned into grouting zones with stoppers.
2. A single grouting zone should be 300 m^2 to 600 m^2.
3. The height of a grouting zone should be 9 m to 15 m.
4. The inlet, outlet and exhaust pipes in a grouting zone should be centrally arranged.

5   The grout stopper shall be connected to the waterstop that also serves as a grout stopper.

6   A repetitive grouting system should be adopted as required.

11.4 Joint Grouting

11.4.1 Before water retaining, joint grouting shall be completed and grout setting shall reach the required strength.

11.4.2 The water retaining plan before closure grouting and the closure measures of an arch dam shall be developed according to the stress analysis for the construction period, and meet the safety requirements in the construction and initial operation periods.

11.4.3 The joint grouting shall meet the following requirements:

1   The concrete on both sides of and above the grouting zone shall reach the design temperature of joint grouting.

2   The joint opening should be no less than 0.5 mm; otherwise, measures shall be adopted.

3   Except for the crest, the thickness of the upper cover layer should be no less than 6 m.

4   The age of the dam concrete on both sides of the joints should be no less than 90 d, and the age of the cover concrete should be no less than 28 d.

11.4.4 Joint grouting shall comply with the current sector standard DL/T 5148, *Technical Specification for Cement Grouting Construction of Hydraulic Structures*, and meet the following requirements:

1   The grouting pressure and joint opening increment shall be both controlled. The grouting pressure shall be restricted to joint opening increment if the joint opening increment has already reached the design value whereas the grouting pressure has not.

2   The grouting pressure at the inlet pipe should be 0.3 MPa to 0.6 MPa, while the grouting pressure at top layer may be properly lowered.

3   The maximum opening increment at the top of grouting zone shall not exceed 0.5 mm.

4   Adjacent grouting zones at the same elevation shall be pressure-equalized using water, and the bottom pressure shall not exceed 0.2 MPa.

11.5 Gallery and Access

11.5.1 Galleries shall be set in arch dams except low dams and thin medium dams.

11.5.2 Galleries and shafts shall exhibit at least one of the following functions:

1 Curtain grouting.

2 Drainage of the dam and foundation.

3 Inspection and maintenance of the dam.

4 Monitoring.

5 Operation of gates; internal lighting; ventilation; and laying of compressed air, water, and power pipelines.

6 Dam cooling, and longitudinal and transverse joints grouting during construction.

7 In-dam transportation and other requirements.

11.5.3 Galleries should avoid the tensile stress zone of the dam. The distance between the gallery and other openings in the dam should be no less than 3 m. The distance from the upstream wall of the longitudinal gallery to the upstream surface of the dam should be 0.05 to 0.10 times the working head at the upstream surface of the dam, and shall not be less than 3 m.

11.5.4 The elevation of galleries shall be comprehensively determined according to the dam height, locations of the dam outlets, and tunnel locations at the dam foundation. Galleries at different elevations should be interconnected by elevator shafts, dam rear bridges, and downstream passages.

11.5.5 Foundation grouting galleries, which also serve the purpose of drainage and inspection, shall be set in high or medium dam. Foundation drainage galleries should be set when the normal downstream water level is high or the dam is thick.

11.5.6 The longitudinal slope of the foundation grouting gallery should be less than 45°. The galleries may be arranged in layers and connected by shafts when the abutment slope is steeper than 45°.

11.5.7 The monitoring galleries may use other galleries or be separately set. Inverted plumb line holes at the monitoring galleries shall not be located at the centerline of the main gallery, and should be located at the branch galleries downstream of the main gallery.

11.5.8 The cross-section geometry of the dam galleries, elevator shafts and sumps shall be determined according to the structural stress conditions and construction methods. The upper part of gallery section may be arc-shaped or trapezoidal. The cross-section of elevator shafts and sumps may be rectangular or square. Galleries should be cast in place, or may be formed by fabricated members.

11.5.9 Cross-sectional dimensions of galleries shall meet the following requirements:

1 Dimensions of foundation grouting galleries shall be determined according to the size and working space of drilling and grouting equipment. The width should be 2.5 m to 4.0 m and the height should be 3.0 m to 4.5 m.

2 Dimensions of monitoring galleries shall be determined according to equipment layout and monitoring requirements.

3 Dimensions of other galleries shall guarantee their functionality and free passage.

11.5.10 Gallery drainage shall meet the following requirements:

1 Drainage ditches shall be set in the galleries.

2 The width and depth of a drainage ditch should be no less than 25 cm.

3 Drainage ditches shall be directed to the downstream or sumps, and the drainage mode should be free drainage for sumps higher than downstream water level and pumping drainage for sumps lower than downstream water level.

4 Pumping drainage shall have backup equipment and backup power supply.

11.5.11 Lighting and ventilation facilities shall be set in galleries, and emergency lighting should also be set.

11.5.12 The dam rear bridges shall meet the following requirements:

1 The elevations of the bridges shall be determined according to the dam height, downstream water level, and dam outlet and gallery arrangement.

2 The bridge width shall meet the requirements for transportation, construction, and patrol inspection during the operation period.

3 The bridge should be of a cantilever or fabricated structure.

4 The bridge shall be able to bear both operating loads and construction loads.

11.5.13 Elevators shall be provided for extra-high dams, should be provided for high dams, and may be provided for medium or low dams. The following requirements shall be met:

1. Elevator shafts shall not affect the structural safety of dams.
2. The bottom stop of elevators should reach the lowest-level gallery.
3. Elevator size and rated loading capacity shall meet the transportation requirements of staff and related equipment.

11.5.14 Other passages on the downstream dam surface shall be designed according to the requirements for inspection, maintenance and transportation. Ladders for gate shafts, valve shafts and gate piers shall be set according to the requirements for installation, inspection and maintenance.

11.5.15 Safety and protections shall meet the following requirements:

1. The entrance and exit of the gallery leading to the outside of the dam shall be provided with protective doors according to the requirements of security, anti-freezing and flood protection.
2. Galleries with slopes exceeding 45° shall be equipped with platforms and handrails in sections.
3. All passages and platforms outside the dam shall be set with guardrails.
4. The design of ladders shall comply with the current national standard GB 4053.1, *Safety Requirements for Fixed Steel Ladders and Platform—Part 1: Steel Vertical Ladders*.
5. Fire protection design shall comply with the current national standard GB 50872, *Code for Fire Protection Design of Hydropower Projects*.

11.6 Waterstops and Drainage

11.6.1 Waterstops shall be set at the following positions of the dam:

1. The upstream side of the transverse joints.
2. The downstream side of the transverse joints below the highest downstream flood level.
3. The transverse joints crossing the overflow surface.
4. The contact surface between the steep-slope monolith and the foundation.
5. The bottom of the transverse joints contacting the foundation surface.

11.6.2 At the upstream side of a transverse joint, 1 or 2 waterstops should be set for a medium or low dam, 2 waterstops shall be set for a high dam, and 2 or 3 waterstops should be set for an extra-high dam.

11.6.3 In addition to the current sector standard DL/T 5215, *Specification for Waterstop for Hydraulic Structures*, the material and structure type of waterstop shall meet the following requirements:

1 Copper waterstops shall be adopted at the following positions:

1) The first waterstop on the upstream side of transverse joints of a medium dam.

2) The first and second waterstops on the upstream side of transverse joints of a high dam.

3) The waterstops on the downstream side of transverse joints below the highest downstream flood level of a high dam.

4) The waterstops in the transverse joints crossing the overflow surface.

5) The waterstops at the contact surface between the steep-slope monolith and the foundation.

2 Plastic or rubber waterstops may be adopted at the following positions:

1) The first waterstop at the upstream side of transverse joints of a low dam.

2) The second waterstop at the upstream side of transverse joints of a medium dam.

3) The third waterstop at the upstream side of transverse joints of an extra-high dam.

4) The waterstops at the downstream side of transverse joints of a low or medium dam.

3 The thickness of copper waterstops should be 1.0 mm to 1.6 mm.

4 Copper waterstops should be formed into the " ⌒ " shape, with the central convex pointing to the seepage direction.

11.6.4 The connection of waterstops shall meet the following requirements:

1 The buried depth of copper waterstops in the concrete on both sides of the transverse joints shall be 20 cm to 30 cm.

2 A waterstop pit with a depth of 30 cm to 50 cm should be set at the bottom of the upstream waterstop of the transverse joints, and anchor rebars should be set in the pit.

3 Waterstops at the contact surface between the steep-slope monolith and

the foundation may be embedded in waterstop ridges or slots along the bedrock on the slope.

4   When multiple waterstops are set at the upstream side of transverse joints, horizontal waterstops should be set as separations with an interval of 1 to 2 times the grouting layering height.

5   Waterstops at the contact surface between the steep-slope monolith and the foundation shall be connected to those of transverse joints.

6   Waterstops on the overflow surface shall be connected or sealed at the intersection with embedded parts of metal structures of the gate sill.

11.6.5 The setup of dam drainage shall depend on the climate zone and the dam thickness. Thin arch dams in mild zones need not be set with dam drainage. The arrangement of the dam drainage shall meet the following requirements:

1   The dam drainage shall be set near the upstream surface of the dam.

2   The dam drainage may be vertical, or horizontal plus vertical.

3   The vertical drainage shall extend upward to the upper gallery, the dam crest or below the overflow surface, and extend downward to the lower gallery, the horizontal drainage pipes, or the header pipe. Sharp elbow should not be arranged at the connection of the dam drainage and the gallery.

4   Horizontal drainage shall be set at concrete lift surface and shall be connected to the vertical drainage.

5   The distance from the drainage to the upstream dam surface should be 0.05 to 0.10 times the working head at the dam surface, and shall not be less than 3 m.

6   A drillhole, drawing pipe or embedded pipe may be adopted for drainage. A perforated pipe, plastic blank pipe, or prefabricated sand-free concrete pipe may be used as the embedded pipe.

7   When only the vertical drainage is set, the vertical drainage spacing should be 2.5 m to 4.5 m. In the presence of horizontal drainage, the vertical drainage spacing shall be determined according to demands.

8   The internal diameter of the drainage pipe should be 15 cm to 20 cm. The diameter of the drilling hole shall not be less than 75 mm.

12 Temperature Control and Crack Prevention Design

12.1 General Requirements

12.1.1 The meteorological and hydrological data used for the design of the temperature control and crack prevention for an arch dam shall meet the following requirements:

1. The mean annual temperature and intra-annual amplitude, mean monthly or 10 d temperature, long-term extreme air temperature, mean daily temperature and daily air temperature amplitude, abrupt air temperature drop amplitude, duration and frequency, and sunshine effect shall be determined according to the observed data from the meteorological station in dam site. When there is no meteorological station at the dam site or the meteorological station has only short data series, the measured data from the meteorological station near the dam site may be collected, and be corrected considering the latitude, longitude, and elevation.

2. The mean annual and mean monthly river water temperature shall be determined based on the measured data of the natural river water temperature from the hydrological station at dam site. In the absence of hydrological station, a simple observation point should be set at the dam site to obtain the river water temperature data; or the measured data from the upstream or downstream hydrological stations and cascade reservoirs, as well as data from local tributaries, should be collected, and adopted after correction.

3. Mean annual and mean monthly ground temperature shall be determined based on the measured data at the dam site. In the absence of measured data, those from neighboring meteorological stations may be used after analysis.

4. Reservoir water temperature distribution should be determined by analogy with similar projects.

12.1.2 The concrete property parameters, used in the design of the temperature control and crack prevention of the arch dam, shall meet the following requirements:

1. Concrete property parameters shall include mechanical, deformation and thermal parameters.

2 The concrete property parameters of a high dam shall be determined by tests.

3 For a medium or a low dam, the concrete property parameters should be determined by tests, and may be taken with reference to the data of similar projects when the test data are not available.

4 The items of the concrete property test shall meet the requirements listed in Article 4.3.8 of this code.

12.1.3 In the temperature field and thermal stress analysis, the finite element method shall be used for an extra-high dam and should be used for a high or medium dam, and the finite difference method or the empirical formula method may be used for a low dam.

12.1.4 The temperature control criteria shall be determined according to the natural conditions at the dam site, concrete property, dam structures and construction methods, which shall be determined by calculation for a high dam, and should be determined by calculation for a medium dam, and may be determined by experience based on similar projects for a low dam.

12.1.5 Temperature control measures shall be determined based on temperature control criteria, concrete property, and construction methods, which shall be determined through calculations and analysis for a high dam.

12.1.6 The temperature control and crack prevention design for the plunge pool and end dam may be performed in accordance with the current sector standard NB/T 35092, *Design Code for Temperature Control of Concrete Dam*.

12.2 Control Criteria for Temperature and Thermal Stress

12.2.1 The control criteria for concrete thermal stress may be determined according to the comprehensive safety factor method or the partial factor method.

12.2.2 When the partial factor method is used, the concrete thermal stress during the construction period shall satisfy the following formula:

$$\gamma_0 \sigma \leq \varepsilon_p E_c / \gamma_{d3} \tag{12.2.2}$$

where

γ_0 is the structural importance factor, which may be taken as 1.10, 1.05, and 1.00 for structures of safety Grades Ⅰ, Ⅱ, and Ⅲ, respectively;

σ is the sum of the thermal stress caused by various temperature differences (MPa);

ε_p is the characteristic value of the concrete ultimate tensile strain;

E_c is the characteristic value of the concrete elastic modulus (MPa);

γ_{d3} is the structural factor for thermal stress control for the serviceability limit state, taken as 1.65.

12.2.3 When the comprehensive safety factor method is used, the concrete thermal stress during the construction period shall satisfy the following formula:

$$\sigma \leq \varepsilon E / K_f \tag{12.2.3}$$

where

ε is the concrete ultimate tensile strain;

E is the concrete elastic modulus (MPa);

K_f is the comprehensive safety factor, which should be taken as 1.5 to 1.8 for a high or medium dam, and 2.0 for an extra-high dam.

12.2.4 The allowable temperature difference of foundation concrete should be determined as per Table 12.2.4, and shall be determined by calculation for the following cases:

1. The linear expansion coefficient of aggregate differs considerably from $1.0 \times 10^{-5}/°C$.

2. The ultimate tensile strain of concrete at 28 d age is lower than 0.85×10^{-4}.

3. The elastic moduli of foundation rock and concrete differ considerably from each other.

4. Tests or measured data demonstrate obvious swelling or shrinking deformation of the concrete.

5. Concrete placement in the foundation restrained zone might experience a long suspension.

6. Concrete blocks have a length-width ratio greater than 2.0.

7. Concrete backfill, concrete plugs and concrete blocks at steep slope.

8. Concrete blocks overflowed by floods during construction period.

9. The temperature of the dam monolith during the construction or operation period might be lower than its steady temperature.

Table 12.2.4 Allowable temperature difference of foundation concrete (°C)

Height above foundation surface h	Length of the longer side of concrete block l				
	Less than 17 m	17 m - 21 m	21 m - 30 m	30 m - 40 m	40 m to full section
Strongly restrained zone $0 - 0.2 l$	26 - 25	25 - 22	22 - 19	19 - 16	16 - 14
Weakly restrained zone $0.2 l - 0.4 l$	28 - 26	26 - 25	25 - 22	22 - 19	19 - 17

NOTES:

1 h starts from the peak of the foundation surface; l denotes the longest length of the foundation horizontal projection.

2 The strongly restrained zone in the dam monolith shall also include concrete between the highest and lowest points on foundation surface.

12.2.5 The allowable temperature difference between the upper and lower concrete lifts should be within 15 °C to 20 °C, whereas that for a high dam shall be determined by calculation.

12.2.6 The allowable temperature difference between the interior and exterior concrete should be determined by thermal stress analysis.

12.2.7 During the construction period, each concrete block shall be placed continuously and evenly, where the placement time difference between adjacent blocks should be less than 28 d, and the height difference between adjacent monoliths should not exceed 12 m, which may be increased after study and demonstration.

12.3 Temperature Control Measures

12.3.1 Lift thickness of concrete placement shall be determined by calculation according to temperature control criteria, placing capability and temperature control measures, and shall meet the following requirements:

1 The lift thickness of concrete in restrained zone should be 1.5 m to 2.0 m.

2 The lift thickness of concrete beyond restrained zone should be 2.0 m to 3.0 m.

3 The lift thickness of concrete over old concrete should be 1.5 m to 2.0 m.

4 Increase in the concrete lift thickness shall be studied and demonstrated.

12.3.2 The time interval between the two lifts should be 5 d to 7 d; if longer than 28 d, the temperature difference shall be controlled within the allowable limit between the upper and lower concrete lifts.

12.3.3 In the mild zone, dam concrete near the foundation should be placed during low temperature months, and in the high temperature season, it should be placed during low temperature times, namely, morning, evening or night.

12.3.4 Construction and protection measures shall be proposed for concrete placement in the high temperature season and severe cold season.

12.3.5 The temperature at mixer outlet and concrete placing temperature shall be determined by temperature control criteria, and may be controlled by the following measures:

  1 Frost resistance or sun resistance of raw materials.
  2 Precooling or preheating of aggregates.
  3 Spraying water.
  4 Concrete mixed with cold or hot water.
  5 Protection during transportation.
  6 Concrete surface water spray or heat retention.

12.3.6 Curing measures shall be taken for exposed concrete surface, and should last till design age when conditions permit.

12.3.7 Surface insulation of concrete shall be designed according to meteorological conditions and allowable temperature difference between the interior and exterior concrete, and shall meet the following requirements:

  1 The time and criteria of surface insulation shall be determined by thermal stress analysis of surface concrete.
  2 Surface insulation methods and insulation materials requirements shall be determined by structure locations and construction conditions.
  3 Thermal insulation materials shall be fireproof and environmentally friendly, and permanent materials shall be durable.
  4 Insulation measures shall be proposed for the concrete surface and transverse joint surfaces to resist cold wave, sudden temperature drop and high temperature hit.

12.3.8 Cooling with embedded water pipes, surface water flow and water impounding, or their combination may be adopted for the concrete cooling after placement. Cooling with water pipes should be taken for dams subjected to requirements of joint grouting.

12.3.9 The water cooling time, cooling rate and cooling amplitude shall meet the temperature control criteria and allowable tensile stress control criteria for concrete at different ages. The cooling process shall meet the requirements of small temperature difference, uniformity and slowness. The cooling should be performed in 2 or 3 stages.

12.3.10 For a dam higher than 100 m or an extra-high dam, intelligent temperature control technique should be used.

13 Water Release Structures and Their Energy Dissipation and Erosion Control

13.1 General Requirements

13.1.1 The design of water release structures and energy dissipation and erosion control shall include:

1 Determination of the type, quantity, outlet dimensions and elevations for the water release structures, and their layout.

2 Discharge capacity and reservoir flood routing.

3 Shape, hydraulic and structural designs of water release structures.

4 Design of the downstream hydraulic connectivity and energy dissipation and erosion control facilities.

5 Atomization effects and protection design.

6 Abrasion resistance and cavitation resistance design.

13.1.2 The grades for water release, energy dissipation structures, and the flood control standard, shall be determined in accordance with the current standards of China GB 50201, *Standard for Flood Control*; and DL 5180, *Classification & Design Safety Standard of Hydropower Projects*.

13.1.3 In cold and severely cold regions, the design of water release and energy dissipation structures shall comply with the current sector standard NB/T 35024, *Design Code for Hydraulic Structures Against Ice and Freezing Action*.

13.1.4 For a high dam, the design of water release structures should be verified through the overall hydraulic model test. For water release and energy dissipation structures of large-sized projects or with complex hydraulic conditions, the hydraulic design and shape design shall be verified through hydraulic model test. Model test shall be performed in the case of severe cavitation and erosion.

13.1.5 Water release and energy dissipation structures shall be accessible for inspection and maintenance.

13.2 Layout of Water Release Structures

13.2.1 Layout of water release structures shall be determined through techno-economic comparison, and shall meet the following requirements:

1 The discharge capacity shall be determined through techno-economic

comparison considering the reservoir regulation capability.

2　The discharge capacity shall meet the requirements of design and check flood release, and emergency drawdown.

3　The water release structure layout and discharge allocation shall be selected according to the discharge magnitude, the discharge impact on the stress and stability of dam, topographical and geological conditions, discharge conditions of dam, and capacity of the downstream channel. The dam release, the bank release, or their combination may be adopted, and the dam release should be given priority.

4　Water release structures shall allow for flexible operation and maintenance.

13.2.2 Release structures of the dam may adopt overflow spillways, surface outlets, middle outlets, deep outlets, bottom outlets, or their combinations. The water release types may adopt dam discharge, dam surface overflow, ski-jump spillway discharge, and roof overflow and jumping over flow of the dam-toe powerhouse.

13.2.3 Types, elevations, quantity, and dimensions of the dam outlets shall be determined considering the following factors:

1　Reservoir operation.

2　Discharge magnitude.

3　Capacity of the downstream channel.

4　Types of energy dissipation.

5　Design and manufacture level of gates.

6　Impacts on dam stress.

7　Impacts on adjacent structures.

8　Effects of construction diversion.

9　Other relevant factors.

13.2.4 Layout of the dam outlets shall consider the safety of the dam and appurtenant structures, and meet the following requirements:

1　The outlets shall not be set at the high stress zone close to the arch dam foundation.

2　Middle outlets, deep outlets and bottom outlets shall be set within the dam monolith, whereas the surface outlet may be set across the transverse joints.

3 Layout of the outlet axis shall meet the requirements of waterway concentration and energy dissipation, and the radial layout or non-radial layout with a small bias should be adopted.

4 Gates at the upstream and downstream ends of the outlets should not occupy the basic section of dam.

5 Multi-outlets in the same layer should be arranged considering the arch shape.

13.2.5 Outflow of the dam outlets shall meet the following requirements:

1 The released water shall smoothly flow into the channel without affecting the operation safety of the arch dam, dam foundation and other structures.

2 The outflow type of the outlets shall be determined according to the water depth of the plunge pool, range of the downstream water area, and atomization effect. For multi-layer outlets with high water head and large discharge, disperse or convergent jets may be adopted.

13.3 Design of Water Release Structures

13.3.1 The shape design of crest overflow weir shall meet the following requirements:

1 The weir surface curve may be designed according to Section E.1 of this code. Other weir surface curves may be adopted after tests.

2 The allowable negative pressure near the crest after atmospheric pressure correction, when the gate is fully opened, shall not be greater than 0.03 MPa at the design flood level, and shall not be greater than 0.06 MPa at the check flood level.

13.3.2 The pressure orifice may be adopted for the middle outlets, deep outlets and bottom outlets. The shape design of a pressure orifice shall meet the following requirements:

1 The longitudinal profile of a pressure orifice may be flat, upturned or downturned according to the exit layout of the pressure segment.

2 The cross-section of the orifice may be rectangular.

3 The top and side shapes of the inlet section may be an ellipse, and the bottom shape may be a parabola, quarter-ellipse or circle arc. Other types of curves may be adopted for the orifice bottom by tests.

4 A pressure transition section should be set upstream of the exit section. The section area of exit shall be smaller than that of the conduit, and the ratio may be 0.88 to 0.93.

  5 The calculation or model test shall be performed to verify the hydraulic pressure line of the pressure section, and no negative pressure shall occur under any operating conditions.

  6 The shape of the sudden-expanded gate slot at exit shall be determined according to the flow regimes and cavitation characteristics at service gate chamber and its downstream section.

13.3.3 The shape of the crest overflow or ski-jump spillway shall take a steeper slope and meet the following requirements:

  1 Cavitation resistance design shall be applied to the surface of flow channel.

  2 Vibration analysis should be performed for flow channel structures with complex hydraulic conditions, and feasible vibration control measures shall be taken.

13.3.4 Hydraulic model tests shall be performed for the roof overflow or jumping over flow of the dam-toe powerhouse, and the following requirements shall be met:

  1 The released flow shall not cause detrimental vibrations that affect the safety of powerhouse structures, operation of equipment and health of staff.

  2 Waterstops and other leakage prevention measures shall be set on the roof of powerhouse.

13.3.5 The design of the diversion bottom outlets shall meet the following requirements:

  1 The temporary flood discharge gap during construction should not be arranged in the monolith with a diversion bottom outlet.

  2 Measures shall be taken to prevent water from entering through the top of the gate slot at the diversion bottom outlet.

13.3.6 The discharge capacity of the water release structures should be calculated in accordance with Section E.2 of this code. For release structures of large-sized projects, or those with complex hydraulic conditions, the discharge capacity shall be verified through hydraulic model tests.

13.3.7 The heights of sidewalls of water release structures shall be in accordance with the current sector standard NB/T 10867, *Code for Design of Spillways*.

13.3.8 The vent for gates of water release structures shall be designed in accordance with the current sector standard NB 35055, *Design Code for Steel Gate in Hydropower Projects*.

13.3.9 The corbel structures at the outlets of the dam body shall meet the requirements of the layout of the flow channels and gate hoists. Short outlet corbel structures should be adopted.

13.3.10 The support structures for the radial gate of the surface outlet shall be determined by the requirements of the structural rigidity of the crest arch, and continuous girders should be adopted. Support structures for radial gates of middle outlets, deep outlets and bottom outlets should be of the deep beam type.

13.3.11 The structural design and reinforcement design of the outlets shall comply with the current sector standard DL/T 5057, *Design Specification for Hydraulic Concrete Structures*. The 3D finite element method should be used in stress analysis of the outlets for a high dam.

13.3.12 The layout of the pre-stressed anchor cable of the gate pier shall meet the following requirements:

　1　The main anchor cables may be loop or straight.

　2　The main anchor cables shall be symmetrically arranged along the thrust centerline.

　3　The main anchor cables should be uniformly distributed.

　4　The anchorage section shall extend into the basic dam section. The anchorage ends shall be staggered and the rebars shall be set for stress dispersion.

　5　The secondary anchor cables supporting the girders or corbels shall be set according to the structural stress conditions.

13.4 Design of Energy Dissipation and Erosion Control

13.4.1 The design of energy dissipation and erosion control shall consider the following factors:

　1　Working head and outlet type.

　2　Downstream topographical and geological conditions.

　3　Water range, depth and water level variation of downstream channel.

4   Requirements of downstream slopes and structures for atomized rain intensity control.

  5   Ship pass, log pass, fish pass, sediment flushing, floating debris pass, drift ice pass, etc.

13.4.2 The selection of the energy dissipation types should meet the following requirements:

  1   For water release structures on dam, ski-jump or free-fall energy dissipation should be adopted.

  2   Hydraulic jump energy dissipation may be adopted for water release structures on a low or medium dam, and should be adopted for a low or medium dam with requirements of downstream steady flow and slightly varied water level, but should not be adopted for a project with ice or floating debris release task.

  3   Submerged bucket energy dissipation may be adopted for a low or medium dam where the tail water is deep and the downstream riverbed and banks have sufficient capability against scouring.

  4   Combined energy dissipation type may be adopted when several types of discharge outlets are set in dam.

13.4.3 The design of ski-jump energy and free-fall energy dissipation shall meet the following requirements:

  1   Continuous, differential, contractive, expansive and other irregular shapes of flip buckets may be adopted for ski-jump and free-fall flow.

  2   The flip bucket elevation shall ensure the free ski-jump.

  3   The lower edge of the free-fall flow shall be fully aerated.

  4   The hydraulic factors of ski-jump energy dissipation should be calculated in accordance with Section E.3 of this code. Hydraulic factors of free-fall energy dissipation should be calculated in accordance with Section E.4 of the code.

  5   The depth and range of the scour pit shall not affect the safety of the arch dam and bank slopes; otherwise, engineering measures such as short apron, end dam, and plunge pool shall be taken according to actual conditions.

13.4.4 The hydraulic jump energy dissipation may be designed in accordance with the current sector standard NB/T 10867, *Code for Design of Spillways*.

13.4.5 Protective range and measures shall be studied for the downstream channel of energy dissipation works according to the outflow velocity and non-scouring velocity of natural channel.

13.5 Plunge Pool and End Dam

13.5.1 The lining and protection of the plunge pool shall be determined by inflow hydraulic conditions, water depth in the plunge pool, and topographical and geological conditions of riverbed. Protection after scouring, pre-excavation without lining, slope protection without bottom protection, or full concrete lining may be adopted.

13.5.2 An end dam should be set at the end of plunge pool when downstream water is not sufficiently deep or maintenance is required. The downstream cofferdam may be modified into an end dam if possible.

13.5.3 The layout of the plunge pool shall be determined according to the length, width, location and estimated dynamic water pressure of jet impact area, which should be verified by hydraulic model test, and shall meet the following requirements:

1. The depth of plunge pool shall meet the following requirements:

    1) The foundation surface elevation of the plunge pool shall be determined according to the hydraulic and geological conditions, and should not be lower than the dam foundation surface.

    2) The inflow water shall form submerged hydraulic jump.

    3) The dynamic hydraulic and fluctuating pressures shall be controlled within the bearing capacity of the pool floor.

2. The length of the plunge pool shall not be less than the sum of the trajectory distance and submerged hydraulic jump length, and the outflow should return to subcritical flow.

3. The width of plunge pool shall be comprehensively determined according to the topographical and geological conditions, width of the inflow nappe, flow regime in the pool, and dynamic hydraulic pressure distribution.

4. The design of the plunge pool sidewall shall meet the following requirements:

    1) A freeboard above design flood level of the plunge pool should be reserved.

    2) The top width of sidewall should meet the access requirements for inspection and maintenance.

3) The sidewall shall be connected with the atomization protection works of the slopes.

13.5.4 The cross-section of the plunge pool may be trapezoidal, compound trapezoidal, or inverted arch-shaped. For trapezoidal or compound trapezoidal sections, the junction of the floor and sidewall should be rounded.

13.5.5 The concrete lining thickness of sidewall and floor of the plunge pool shall be determined according to geological and hydraulic conditions. The block sizes of the concrete lining shall be determined according to the requirements of floating-resistant stability and temperature control. Permanent joints of the concrete lining should be of contact joints.

13.5.6 The floating-resistant stability of the plunge pool lining should be calculated in accordance with the current sector standard NB/T 10867, *Code for Design of Spillways*. Anchors or anchor bundles should be set at concrete floor and sidewall, and be connected with surface rebar of the plunge pool.

13.5.7 The concrete of the plunge pool floor should be placed continuously and evenly. Surface roughing and bonding measures shall be taken for lifts.

13.5.8 The structural reinforcement of the plunge pool should be calculated by the finite element method or beam-on-elastic-foundation method, and reinforcements shall be embedded in the abrasion-resistant layer.

13.5.9 The parapet wall with gaps should be set on top of plunge pool sidewalls, and railings shall be set at gap location.

13.5.10 The section of end dam shall be designed according to operation and maintenance conditions of the plunge pool. The height shall be determined according to requirements of water depth in the plunge pool, no hydraulic jump at downstream of the end dam, and maintenance.

13.5.11 The crest width of end dam shall meet the requirements of plunge pool maintenance, and the junction between crest and downstream dam slope should be rounded.

13.5.12 Gravity-type end dam shall be designed in accordance with the current sector standard NB/T 35026, *Design Code for Concrete Gravity Dams*.

13.5.13 When the drainage and pumping facilities are set in a plunge pool, grout curtain and drainage curtain shall be set at the end dam.

13.5.14 Drainage and pumping facilities should be provided for a plunge pool whose lining withstands high uplift pressure, and shall meet the following requirements:

1 A drainage and pumping system may consist of grout curtain, drainage holes, drainage trench, drainage gallery, sump, pumping house, etc.

2 Waterstops shall be set in the structural joints of concrete lining. Drainage trench should be set at the bottom of structural joints. The bottom of drainage trench shall be bedrocks, in which drainage holes should be set connecting the trench. Drainage trench shall be connected with drainage gallery, through which the flow shall be discharged into sump.

3 The slopes of trench in drainage galleries shall ensure free flow by gravity to sump.

4 Galleries for curtain grouting and drainage of the end dam should be interconnected with drainage galleries of the plunge pool.

5 Sump and pumping house may be set inside the mountains.

13.5.15 Inspection and maintenance design of the plunge pool should meet the following requirements:

1 The pumping system and water filling system should be set for the plunge pool.

2 Sump should be set at the end of the plunge pool floor.

3 Inspection and maintenance access to the plunge pool should be set, and should be set at the arch dam toe or the junction of the end dam and the plunge pool sidewalls.

13.5.16 An apron should be set downstream of the end dam according to scouring conditions of the outflow on dam foundation.

13.6 Design of Protection Against Cavitation and Abrasion

13.6.1 The hydraulic design of water release and energy dissipation structures shall focus on the following parts or zones prone to damage by cavitation:

1 Pier head, gate slots, transition section from curve to straight, the ogee and nearby section, bend section, flared section or slope change section, and abrupt changes in the flow boundary.

2 Special-shaped buckets, slotted buckets and dispersing piers.

3 Baffle piers and toe piers.

4 Areas where the flow velocity is over 20 m/s, or the flow cavitation number is lower than 0.30.

13.6.2 The flow cavitation number in the high velocity flow area should be greater than the initial cavitation number. The flow cavitation number may be calculated according to Section E.5 of this code.

13.6.3 For the positions and areas prone to damage by cavitation, the cavitation resistance design shall meet the following requirements:

1   The shape design shall comply with Sections 13.3 to 13.5 of this code.

2   The surface local unevenness shall comply with Table 13.6.3.

3   The areas where the flow cavitation number is lower than 0.30 or the flow velocity is over 35 m/s shall be set with reasonable aeration facilities.

4   Materials with adequate cavitation resistance shall be adopted for surface protection.

5   The gates shall be operated in a reasonable way including opening sequence and gate combination, and the openings.

Table 13.6.3 Local surface unevenness

Flow cavitation number σ		>0.60	0.60 - 0.35	0.35 - 0.30	0.30 - 0.20		0.20 - 0.15		0.15 - 0.10		<0.10
Aeration facilities		–	–	–	No	Yes	No	Yes	No	Yes	Yes
Protrusion height control (mm)		≤25	≤12	≤8	<6	<15	<3	<10	Modify design	<6	Modify design
Treated slope	Upstream	–	1/10	1/30	1/40	1/8	1/50	1/10	Modify design	1/10	Modify design
	Downstream	–	1/5	1/10	1/10	1/4	1/20	1/5		1/8	
	Side	–	1/2	1/3	1/5	1/3	1/10	1/3		1/4	

NOTE The protrusion height control should be checked using a 2 m guiding rule.

13.6.4 The abrasion and cavitation resistance materials of water release structures on a sediment-laden river shall be selected considering the combined actions of silt-carrying flow abrasion, bed load impact and cavitation.

13.6.5 The abrasion and cavitation resistance design of dam outlets shall meet the following requirements:

1   The concrete of outlet wall shall have adequate abrasion and cavitation resistance.

2   Steel lining should be adopted for the walls of middle, deep and bottom outlets for discharging or emptying with high flow velocity or high

pressure, and shall be solidly bonded to dam concrete.

13.6.6 Surface concrete materials for the abrasion and cavitation resistance may be selected in accordance with the current sector standard DL/T 5207, *Technical Specification for Abrasion and Cavitation Resistance of Concrete in Hydraulic Structures*. Concrete zoning should be designed according to cavitation number, velocity and sediment concentration of water flow.

13.6.7 The strength class difference between abrasion and cavitation resistance concrete in surface layer and adjacent conventional concrete should not exceed 10 MPa, and the concrete shall be placed continuously in the same lift.

13.7 Protection for Flood Discharge Atomization Area

13.7.1 Flood standard of the flood discharge atomization area should be consistent with that of the energy dissipation and erosion control structures.

13.7.2 The intensity and range of the atomization-induced rainfall may be determined according to the experience of similar projects. For large-sized projects, or projects with great hazard from flood discharge atomization, the intensity and range of the rainfall should be determined by numerical simulation or hydraulic model test.

13.7.3 The flood discharge atomization area may be classified as area of water nappe and splash, area of heavy fog and rainstorm, area of thin fog and rainfall, and area of thin fog and water vapor dispersion.

13.7.4 Facilities including switchyard, power station circuit outlets, high-voltage wires, ground powerhouse and traffic, etc. should not be set in water nappe and splash area and heavy fog rainstorm area. If inevitable, effective protective measures shall be taken.

13.7.5 Protective measures for slopes in flood discharge atomization area shall be comprehensively determined by atomization rainfall intensity, local natural rainfall intensity and frequency, slope geological condition, slope grade and stability analysis results. Measures such as concrete slope protection, shotcrete slope protection and systematic drainage may be adopted.

14 Safety Monitoring Design

14.1 General Requirements

14.1.1 The safety monitoring system shall be designed in accordance with the current sector standard DL/T 5178, *Technical Specification for Concrete Dam Safety Monitoring* based on the structure grade, dam height, type and characteristics of structure, and topographical and geological conditions, and shall be able to monitor the project life-cycle performance and safety, and to provide guidance for the construction and feedback to the design.

14.1.2 The layout of safety monitoring facilities shall meet the following requirements:

1. The monitoring layout shall be targeted.

2. The monitoring sections and positions shall be representative, and the keypoints shall be highlighted.

3. Important monitoring items should use two or more monitoring methods.

4. The monitoring instruments for important physical quantities should be arranged with backup.

5. The monitoring equipment and instruments shall have the required measuring range, precision and long-term reliability in a harsh environment.

6. The important items should be monitored automatically, and the main items for a high dam shall be monitored online.

14.1.3 The safety monitoring design shall meet the following requirements:

1. The permanent patrol inspection accesses and the monitoring galleries, shafts and caverns shall use the dam galleries as much as possible, and separate accesses shall be set when necessary.

2. The monitoring station shall have appropriate accessibility, lighting, and protection against moisture, humidity, wind and frost.

3. The cable shall be protected by cable conduits and trays.

4. The appearance of the monitoring facilities and equipment shall meet the requirements for reaching the standard and putting into production.

5. Important monitoring equipment shall be provided with necessary anti-theft and security protection.

6　The monitoring instruments should be installed using reserved pits, embedded parts or drillholes.

7　The monitoring requirements for initial impoundment shall be proposed before impoundment.

8　During the construction and initial impoundment periods, the monitoring data shall be processed and analyzed. Necessary temporary monitoring measures shall be taken when the permanent monitoring facilities are not ready before the first impoundment. The monitoring data shall be compiled in accordance with the current sector standard DL/T 5209, *Specification of Information Compilation for Concrete Dam Safety Monitoring*.

9　For a high dam, safety monitoring indicators of displacement, seepage pressure and seepage quantity at key parts during the initial impoundment should be proposed.

14.2　Monitoring Items

14.2.1　The patrol inspection items shall be in accordance with the current sector standard DL/T 5178, *Technical Specification for Concrete Dam Safety Monitoring*.

14.2.2　The classification and selection of monitoring items shall be based on the dam height in accordance with Appendix F of this code.

14.2.3　The special monitoring design shall be determined according to the structure grade and the project-specific situation, and may include:

1　Stability of slopes near the dam.

2　Stability of reservoir bank landslide.

3　Reservoir-induced earthquake.

4　Seismic response of the dam.

5　Hydraulic characteristics of water release and energy dissipation structures, and gate dynamic response.

6　Stress conditions of anchorage structure.

7　Stability of surrounding rocks of underground cavern.

8　Monitoring of foundation surface relaxation deformation due to excavation.

9　Control network for deformation monitoring.

10 Safety monitoring automation system.

11 Temperature control during concrete construction period and grouting control.

12 Settlement monitoring of reservoir basin.

13 Forward and feedback analyses of safety monitoring during initial impoundment period.

14 Monitoring automation for an extra-high dam in construction.

14.3 Monitoring Instrument Layout

14.3.1 The layout of deformation monitoring sections for dam and foundation shall meet the following requirements:

1 The cantilever monitoring sections should be equidistantly and symmetrically arranged, and monoliths at the crown cantilever and a quarter of the arch length shall be given priority. The number of cantilever monitoring sections should not be less than 5 for a high dam, and not be less than 7 for an extra-high dam. The cantilever monitoring section should extend into bedrock by 1/2 to 2/3 of the dam height.

2 The arch monitoring sections should be equidistantly arranged, and those at the crest, 2/3 and 1/3 of the dam height shall be selected in priority. The number of arch monitoring sections should not be less than 4 for a high dam, and not be less than 6 for an extra-high dam. The arch monitoring section shall extend into the foundation.

3 The vertical monitoring section shall be set at the abutment of the crest arch.

14.3.2 The arrangement of deformation measuring points of dam and foundation shall meet the following requirements:

1 Deformation measuring points shall be arranged at the intersection of an arch monitoring section and a vertical monitoring section.

2 Horizontal displacement measuring points shall be arranged at the dam crest, dam surface, galleries or adits.

3 Vertical displacement measuring points shall be arranged in the middle of each monolith.

4 Inclination measuring points shall be arranged in the branch gallery along cantilever.

5 The measuring line for chord should be arranged at the dam crest and

abutments where the arch thrusts are large.

6 Both the plumb line and surface measuring points shall be arranged for a high dam.

14.3.3 The arrangement of measuring points for the transverse joints opening shall meet the following requirements:

1. The measuring points shall be set at the transverse joints in the monoliths selected for deformation monitoring.

2. The measuring points shall be located in the middle elevation of the grouting zone, and their number from upstream to downstream at the same elevation should be 2 to 3.

3. The measuring points shall be set in the transverse joints in grouting zones within the strongly restrained zone.

4. The measuring points shall be set in all the transverse joints of grouting zones at the crest of a high dam.

14.3.4 The arrangement of measuring points for openings of structural joints and cracks shall meet the following requirements:

1. The measuring points shall be set at the dam and foundation contact joints, perimeter joints, short joints at dam heel, and plug contact joints.

2. The measuring points of the dam and foundation contact joints shall be arranged at the heel and toe of the deformation monitoring monolith, and should be arranged at each monolith for a high dam.

3. The measuring points for cracks during construction period shall be arranged according to the crack size and the influence on the structure.

14.3.5 The arrangement of deformation monitoring of dam foundation and near-dam slopes shall meet the following requirements:

1. Surface deformations shall be monitored for near-dam slopes, and deep deformation should be monitored for the slopes with complex geological conditions.

2. For the dam foundation with complex geological conditions, the deformation of rock mass shall be monitored. The measuring points should be arranged at the dam heel and dam toe, and the measuring direction should be the same with the main force direction at the part.

3. Foundation with high crustal stresses should be arranged with monitoring points for unloading deformation of rocks.

4 Deformation measuring points shall be arranged at abutments with thin topography, salient isolated peak or large scale treatment for geological defects. The measuring points should be arranged at the part which bears large arch thrust and the measuring direction should be the same as the main deformation direction at this part.

14.3.6 The valley deformation should be measured for a high dam over 100 m or an arch dam with complex geological conditions on the bank slopes in the project area, and shall be arranged according to the following requirements:

1 The position of the measuring line should be arranged in front of the dam, at the dam axis and behind the dam.

2 The measuring points should be arranged in pairs perpendicular to the stream direction at the same elevation on both banks.

3 The endpoints of the measuring line shall be arranged on stable rocks, and should be set in caverns.

4 Coordinate measurement of the surface measuring points shall be conducted not less than once every six months.

14.3.7 For an extra-high dam, the settlement measuring of the reservoir basin should be performed according to the reservoir capacity and reservoir geological conditions.

14.3.8 The seepage measuring arrangement shall meet the following requirements:

1 The transversal section of the dam foundation seepage measuring shall be arranged parallel to the centerlines of grout curtain and drainage curtain, and the longitudinal section should be set in the deformation measuring monolith.

2 The longitudinal measuring line of dam foundation should be set with 2 to 4 seepage pressure measuring points.

3 The measuring sections for bypass seepage should be set 2 to 3 for both banks respectively, and the measuring points of each monitoring section shall not be less than 3. The measuring holes shall extend below the underground water level before dam construction and should extend into the strong permeable layer.

4 The contact joints of plug should be set with seepage pressure measuring points.

5 The seepage pressure measuring points for the cracks occurring in the construction period and for the geological defects in dam foundation

shall be comprehensively determined according to the location, scale and influence.

6   The leakage measuring points shall be arranged in different zones and segments according to the leakage collection and drainage conditions.

14.3.9 The stress and strain measuring arrangement shall meet the following requirements:

1   The stress and strain measuring arrangement of dam shall meet the following requirements:

   1) The stress and strain measuring section of the dam shall be consistent with the deformation monitoring section.

   2) Three measuring points, namely, the upstream, middle and downstream points, should be arranged at each monitoring line for the dam, and the number should be more than 3 in presence of longitudinal joints.

   3) The plane stress measuring points shall be arranged parallel to the dam surface in the near-surface area, and the spatial stress measuring points should be arranged in the middle area.

   4) The distance from the measuring points to the dam surface shall not be less than 1 m, and to the foundation surface shall be more than 3 m.

   5) The compressive stress meter along cantilever direction or arch direction should be arranged in the strongly restrained zone near the foundation, and shall be arranged in groups with the spatial stress measuring points.

   6) For an extra-height dam, stress measuring points shall be arranged near the dam surface, dam foundation and abrupt changes of foundation surface or other parts with local stress concentration.

2   The stress and strain measuring arrangement for important structures of dam shall meet the following requirements:

   1) The stress measuring points shall be arranged at the outlet of an extra-high dam.

   2) Reinforcement meters shall be arranged for reinforced concrete at dam outlets or piers.

   3) The pre-stressed structure shall be equipped with load cells.

3   The dam foundation stress and strain measuring arrangement shall meet

the following requirements:

1) Micro strain measuring holes may be arranged where the stresses of dam foundation rock mass are high.

2) Anchor stress meters shall be arranged for anchor rods of dam foundation.

3) The pre-stressed anchor cables of dam foundation shall be equipped with load cells.

14.3.10 The temperature measuring arrangement of dam and foundation shall meet the following requirements:

1 The temperature monitoring section should be consistent with the deformation monitoring section.

2 The temperature measuring points shall be arranged by grid according to the distribution of temperature field, and the number of measuring points along the thickness direction of arch shall not be less than 3. The points should be densified at the dam surface and outlets with large temperature gradient.

3 The dam surface temperature measuring points should be 5 cm to 10 cm away from the dam surface.

4 The temperature measuring points of dam foundation should be set for a high dam.

15 Construction Requirements

15.1 Foundation Surface Excavation

15.1.1 The construction sequence of foundation surface excavation and slope support shall be determined according to the structure layout, topography and geology.

15.1.2 The excavation timing and sequence of foundation surface defects shall be comprehensively determined according to the scale and property of the defects, replacement structure type, and upstream and downstream slopes.

15.1.3 Design requirements shall be put forward for the blasting control of foundation excavation of an arch dam according to the dam scale and the rock mass conditions of dam foundation. For a high dam, the parameters concerning blasting control shall be determined by test, and be adjusted during construction according to the monitoring and inspection results.

15.2 Foundation Treatment

15.2.1 The loose and weak rocks on the foundation surface, and sheet-like or sharp-angled rock mass shall be removed.

15.2.2 For the excavation of the replacement holes, wells and grids in weak zones, geological prediction, advance support, and tracking adjustment should be conducted successively.

15.2.3 The drilling of the consolidation grouting holes on the concrete surface shall not break cooling pipes or monitoring cables. The grouting operations shall be zoned and sequenced, and the concentrated grouting for hole groups shall be avoided.

15.2.4 The construction conditions of curtain grouting shall meet the following requirements:

1. The strength of the dam concrete above the grouting zone shall reach 50 % of the design strength or be above 10 MPa.

2. The bedrock consolidation grouting, transverse joint grouting, and contact grouting of relevant positions shall be completed and accepted.

3. The concrete lining, backfill grouting, and consolidation grouting of the surrounding rock of relevant positions shall be completed and accepted.

4. Other construction operations which might affect the quality of curtain grouting shall be completed.

15.2.5 The construction conditions of contact grouting shall meet the following requirements:

1. The weighted concrete cover above the grouting zone shall be sufficiently thick to overcome the grouting uplift.

2. The temperature of replacement concrete in dam and foundation shall meet the design requirements.

15.2.6 The drainage hole drilling shall not commence until the consolidation grouting, contact grouting, and curtain grouting of relevant positions are completed and accepted.

15.3 Concrete Construction

15.3.1 The upstream dam face at the abutment shall be separated from the upstream slope of the abutment. The downstream fillet concrete at the abutment shall be poured integrally with the dam concrete, and the downstream slope concrete may be poured separately from the dam concrete.

15.3.2 The dam foundation replacement concrete poured separately from the dam concrete shall be aged sufficiently and reach the design temperature when covered by the upper dam concrete.

15.3.3 The structural safety of corbels shall be checked for each lift during construction, and auxiliary construction measures shall be taken as required.

15.3.4 The concrete construction quality shall be tracked and inspected throughout the process, and shall comply with the current sector standard DL/T 5144, *Specifications for Hydraulic Concrete Construction*. For concrete cracks found, the following work shall be conducted and treatment shall be done if necessary:

1. Analyze the causes and development trend of cracks according to their locations, sizes, shapes and occurrences.

2. Take necessary measures against crack development according to the locations and effects of cracks, including:

    1) Drill stress release holes at the crack tips.

    2) Set rebars where cracking occurs on the concrete surface.

3. According to the crack locations, sizes, and effects on the structure, carry out dedicated treatment and monitoring, enhance the seepage control of cracks on the upstream dam face, and take measures to restore the integrity of structure subject to through cracking.

15.3.5 The distance from the concrete lift surface to the gallery top should be no less than 1.5 m. If it has to be less than 1.5 m, the concrete lift surface should be connected to the top of gallery walls with a slope of 1 : 1.0 to 1 : 1.5.

15.3.6 Cooling pipes, grouting pipes, construction holes and openings, and quality inspection holes shall be plugged after the intended function is completed.

15.4 Monitoring

15.4.1 For a high dam, the following items during construction period should be monitored considering the construction measures and the commissioning of safety monitoring system:

1. Relaxation of the foundation surface.
2. Vertical displacement of the galleries at different elevations.
3. Seepage quantity before foundation pit filling.
4. Horizontal displacement of arch dam.
5. Concrete temperature during construction period.
6. Joint opening during joint grouting.
7. Uplift due to grouting.

15.4.2 The construction sequence of the monitoring system shall meet the following requirements:

1. The control network for monitoring the deformation of a high dam shall be completed before slope excavation.
2. Multi-point extensometers in the dam foundation shall be installed and accepted before corresponding parts are grouted.
3. The construction of the dam foundation inverted plumb line holes and seepage pressure monitoring facilities shall not commence until the adjacent grouting terminates, and shall be as required by the impoundment plan.
4. The plumb line system shall be installed and tested upon the formation of upper galleries.

15.4.3 The normal plumb line hole should be formed by prefabricated pipe or draft.

16 Initial Impoundment, Operation and Maintenance

16.1 General Requirements

16.1.1 The initial impoundment, operation and maintenance of an arch dam shall include:

1 Requirements for physical progress of impoundment works and structure performance.

2 Requirements for impoundment control.

3 Requirements for structure operation.

4 Requirements for monitoring, and data analysis and evaluation.

16.1.2 The operation and maintenance of the dam safety monitoring system during the operation period shall comply with the current sector standard DL/T 1558, *Code for Operation and Maintenance of Dam Safety Monitoring System*.

16.1.3 The safety evaluation during the operation period shall comply with the current sector standard DL/T 5313, *Guide for Safety Assessment of Large Dams for Hydropower Station in Operation*.

16.1.4 The impoundment process shall meet the requirements of ecological environment protection, reservoir clearing, resettlement, and overall stability of near-dam banks.

16.1.5 The dam shall be in normal performance during impoundment.

16.2 Impoundment and Early-Stage Operation

16.2.1 The physical progress of structures before impoundment shall meet the following requirements:

1 The closure grouting elevation should be higher than the target staged impoundment level.

2 The treatment of the load-bearing area of the dam foundation subject to water pressure shall be completed and accepted.

3 The curtain grouting elevation should not be lower than the impoundment level, and the completed grout curtain below the impoundment level shall be accepted. When the curtain grouting elevation is lower than the impoundment level, the rationality of the impoundment level and the curtain grouting plan shall be demonstrated according to the hydrogeological conditions, seepage path, and construction conditions.

4 The plugging of the construction adits, exploratory adits, and drillholes

below the impoundment level shall be completed and accepted.

- 5 The treatment of the construction defects below the impoundment level shall be completed and accepted.
- 6 The construction of the water release, intake, and energy dissipation structures operating below the impoundment level shall be completed and accepted.
- 7 Treatment of the slopes affected by flood discharge atomization shall be completed and accepted.
- 8 The construction of monitoring facilities related to impoundment shall be completed and accepted.

16.2.2 The impoundment control shall meet the following requirements:

- 1 The rising rate of water level shall be studied and determined according to the project scale, dam structure, valley shape and geological conditions of dam foundation. The rising rate of water level for the upper 1/3 of a high dam should be limited between 2 m/d and 5 m/d.
- 2 The water level for the upper 1/3 of a high dam should be raised in stages, the water level rise at each stage should not exceed 20 m, and the water level holding period should be 5 d to 10 d.
- 3 The monitored deformations, stresses and seepage of dam shall be normal and consistent with general laws.
- 4 The impoundment at subsequent stages shall be determined through analysis and evaluation of monitoring results.

16.2.3 The dam outlets operating during the initial impoundment period shall meet the requirements for energy dissipation and vibration suppression.

16.2.4 Monitoring feedback analysis should be conducted during the initial impoundment period for a high dam and an arch dam with complex geological conditions.

16.2.5 During the early-stage operation of an arch dam, the following work should be performed:

- 1 Study the correlation between the stability of slopes and landslide in reservoir area and the rising-falling rate of water level.
- 2 Conduct prototype observation tests of hydraulics and gate dynamics for a high dam or a project with large flood discharge capacity.
- 3 Conduct comprehensive analysis of monitoring data and evaluation

of arch dam performance according to the loading-unloading cycle of reservoir water.

16.3 Operation and Maintenance

16.3.1 The technical requirements for operation shall be determined based on the project features and the early-stage performance.

16.3.2 Comprehensive analysis of monitoring data and safety evaluation of arch dam operation shall be performed regularly during the operation period.

16.3.3 The severe defects found in patrol inspection and the anomalies indicated by monitoring data shall be tested, analyzed, studied and corrected.

16.3.4 The overflow surfaces, gate pier surfaces, gate slots, gates and hoists of water release structures shall be inspected before and after flood season, and when any anomaly is found during operation, and the potential hazards shall be eliminated before use.

16.3.5 The inspection and maintenance of the plunge pool should be conducted once a year during the early-stage operation period, and may be once every 3 to 5 years during the operation period.

16.3.6 The inspection and maintenance of drainage holes, drainage ditches and drainage pumps for the dam, dam foundation and plunge pool shall be performed once a year as a minimum. When the precipitates heap up unfavorably or show abnormalities, analysis and evaluation shall be conducted and necessary treatment shall be implemented.

16.3.7 The continual observation and inspection shall be performed for the slopes and landslides in the project area and the near-dam reservoir area, and the stability shall be reviewed and countermeasures proposed when necessary.

16.3.8 Special inspection and evaluation shall be carried out during the operation period in the following cases:

1 Flood or pool level exceeding the standard.

2 Extraordinary rainstorm.

3 Earthquake with a seismic intensity of Ⅵ or above.

4 The rising and falling rates of pool level exceeding the operation technical requirements.

5 The pool level lower than the dead water level.

16.4 Analysis and Evaluation of Dam Performance

16.4.1 The safety and performance of an arch dam shall be analyzed and evaluated during the operation period.

16.4.2 The analysis items for the performance of a high arch dam shall be determined according to monitoring results, geological conditions, environmental factors, project scale, structural characteristics, and operation time, and should include:

1 Analysis of the temporal and spatial variation law, periodicity and correlation of deformation, stress, seepage and temperature based on the monitoring results.

2 Monitoring feedback analysis of a high dam or an arch dam with complex geological conditions.

3 Analysis of the monitored displacement, seepage pressure and seepage quantity at key positions.

4 Comparison of the monitoring results and safety monitoring indicators, and analysis of anomalies.

16.4.3 The performance of a high arch dam shall be comprehensively evaluated based on the analysis of monitoring data, considering the correlation between dam deformation and pool level, and the regularity, periodicity and stability of the deformation, as well as monitoring results of seepage quantity and seepage pressure.

16.4.4 When obvious deformation of the reservoir basin or the valley occurs, its impact on the dam safety shall be analyzed.

Appendix A Uplift Calculation

A.1 Uplift Acting on Dam Base

A.1.1 For an arch dam with grout curtains and drainage holes set in dam foundation, the distribution of uplift pressure (Figure A.1.1) shall be calculated as follows according to the layout of grout curtain and drainage holes:

1. When the grout curtain holes and drainage holes are respectively arranged in galleries with a certain distance [Figure A.1.1(a)], the uplift pressure shall be the upstream water depth H_1 at dam heal, and the downstream water depth H_2 at dam toe. The upstream-downstream water level difference H shall equal $H_1 - H_2$, and the uplift pressure shall be $H_2 + \alpha_1 H$ at the curtain centerline, $H_2 + \alpha_2 H$ at the drainage line. The adjacent points shall be connected with a straight line. The values of α_1 and α_2 shall be determined according to the geological conditions, layout of the grout curtain and drainage facilities, and should take 0.40 to 0.60 and 0.20 to 0.35, respectively.

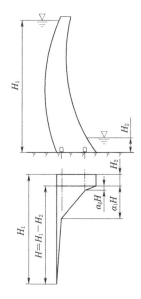
(a) With downstream drainage

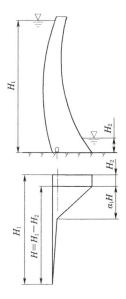
(b) Without downstream drainage

Key
H upstream-downstream water level difference
H_1 upstream water depth
H_2 downstream water depth
α_1 reduction factor for uplift pressure at grout curtain
α_2 reduction factor for uplift pressure at drainage holes

Figure A.1.1 Distribution of uplift pressure

2   When the primary drainage holes are arranged immediately downstream of the grout curtain in the same gallery and a drainage gallery serving as the secondary drainage is arranged between the primary drainage and dam toe, forming a primary-secondary drainage system, the uplift pressure distribution shall be as shown in Figure A.1.1(a). The values of α_1 (at centerline of the gallery) and α_2 should be determined according to the geological conditions, grout curtain and drainage facilities, and should take 0.25 to 0.40 and 0.10 to 0.20, respectively.

3   When the dam is thin and a single row of drainage holes is arranged immediately downstream the grout curtain, the uplift pressure distribution shall be as shown in Figure A.1.1(b). The value of α_1 should take 0.25 to 0.40.

A.1.2 When the geological conditions are suitable and only drainage holes are provided without grout curtain, the distribution of uplift pressure shall be as shown in Figure A.1.1(b), and the value of α_1 shall be properly increased, which should be taken as 0.30 to 0.45.

A.2   Uplift Acting on Abutments

A.2.1 The uplift pressure on the abutment at upstream boundary of the sliding block shall be H_1, and the uplift pressures at the curtain line and drainage lines shall comply with Articles A.1.1 and A.1.2 of this code under various layouts of curtain and drainage.

A.2.2 Reduction factors α_1 and α_2 respectively at curtain and drainage holes should be properly higher than those given in Articles A.1.1 and A.1.2 of this code.

A.2.3 The zero value point of downstream seepage pressure may be assumed at the position 2 to 3 times the arch abutment width downstream of the upstream point of the arch abutment, according to the absolute dam thickness and drainage facilities on the abutment, linearly varying between control points.

A.2.4 The uplift pressure on the abutments in case of drainage failure may be assumed to linearly vary from H_1 to H_2.

A.2.5 The uplift pressure on the abutments shall be determined by three-dimensional seepage computation or test for a high dam of Grade 1 or 2 and an arch dam with complex geological conditions.

A.3   Uplift in Dam

A.3.1 The distribution of the uplift pressure (Figure A.3.1) in a thick or medium thick arch dam shall be determined as follows according to the arrangement of drainage pipes:

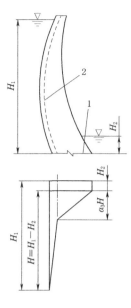

Key

- H upstream-downstream water level difference
- H_1 upstream water depth
- H_2 downstream water depth
- α_3 reduction factor for uplift pressure in dam
- 1   calculation section
- 2   drainage centerline

Figure A.3.1 Distribution of the uplift pressure in a thick or medium thick arch dam

1. The uplift pressure shall be the water depth H_1 above the calculation section at upstream surface, $H_2 + \alpha_3(H_1 - H_2)$ at the drainage centerline, and H_2 at downstream surface, with adjacent points connected by a straight line. The value of α_3 shall be determined according to the concrete quality, and should take 0.15 to 0.30. H_2 shall be zero when the calculation section is above the downstream water level.

2. The uplift pressures at the upstream and downstream surfaces shall be H_1 and H_2, respectively, with a straight line connected when no drainage pipes are set in dam.

Appendix B Single Safety Factor Design for Arch Dam Stress by Trial-Load Method Under Non-seismic Loading

B.0.1 The loads and load values of a concrete arch dam shall comply with Section 5.1 of this code. The load combinations shall include the fundamental combination and special combination. The fundamental combination shall be determined according to persistent and transient situations, and the special combination shall be determined according to the non-seismic accidental situation, as presented in Table 5.2.2 of this code.

B.0.2 The principal compressive stresses calculated using the trial-load method shall not be greater than the allowable compressive stress of concrete which shall be calculated by dividing the strength class of concrete by the safety factor for compressive strength. The safety factor for compressive strength shall be taken from Table B.0.2.

Table B.0.2 Safety factors for the compressive strength

Load combinations		Structure grade		
		1	2, 3	4, 5
Fundamental combinations	Persistent condition	4.4	4.0	3.6
	Transient condition	4.2	3.8	3.4
Special combinations	Accidental condition	3.7	3.4	3.0

B.0.3 The tensile stress control for an arch dam shall meet the following requirements:

1. For a persistent condition, the maximum tensile stress by the trial-load method shall not be greater than 1.2 MPa.

2. For a transient condition, the maximum tensile stress in an unclosed dam monolith should not be greater than 0.5 MPa.

3. For a non-seismic accidental condition, the maximum tensile stress should not be greater than 1.5 MPa.

4. If the tensile stresses in local areas of the dam surface fail to meet the requirements, the overall stability analysis method shall be adopted to evaluate the yield zone and its effect on the dam.

Appendix C Sliding Stability Analysis of Abutment

C.1 Boundaries of a Sliding Block

C.1.1 The boundaries of a sliding block shall include the upstream tension cracking plane, side sliding (cracking) plane, bottom sliding plane and, if necessary, downstream boundary.

C.1.2 A specific discontinuity across the abutments should be selected as the upstream tension cracking plane. When there is no specific discontinuity at the abutment, a hypothetic boundary shall be selected for the upstream tension cracking plane. A curved surface starting from the upstream boundary line of dam foundation surface and extending vertically into the mountain should be selected as the hypothetic boundary.

C.1.3 A side sliding plane should be made up of one single or a set of weak zone(s) and fissure(s) with large dipping angles.

C.1.4 A bottom sliding plane should be made up of one single or a set of weak zone(s) and fissure(s) with small dipping angles.

C.1.5 A free face or other weak zone as well as fissures providing space for the block to slide and deform shall be selected as the downstream boundary.

C.2 Sliding Block

C.2.1 A sliding block shall be determined in the following steps:

1. Analyze the discontinuities within the abutment range and single out those probably affecting the abutment stability.

2. Group the discontinuities into steep dip and gentle dip planes according to their attitudes and spatial locations.

3. Determine the sliding blocks by trial combination and calculation.

C.2.2 The following combinations should be included, and other possible patterns of sliding blocks should be supplemented according to actual situations.

1. Sliding block formed by a steep side sliding plane and a gentle bottom sliding plane.

2. Sliding block formed by a steep side sliding plane, a steep downstream boundary and a gentle bottom sliding plane.

3. Stepped sliding block formed by several same-occurred steep side

sliding planes and several same-occurred gentle bottom sliding planes.

C.3 Shear Strength Parameters of the Sliding Plane

C.3.1 Shear strength parameters of the upstream cracking surface should be taken as zero.

C.3.2 Shear strength parameters of the other sliding planes should meet the following requirements:

1. When the sliding plane is along a weak zone, its shear strength parameters should take the values of this discontinuity.

2. When the sliding plane is along fissures, its shear parameters should be calculated by comprehensive weighting after considering the percentages of rocks penetrated by this sliding plane and connectivity of the fissures.

C.4 Actions on a Sliding Block

C.4.1 The thrust of an arch dam shall be calculated in the following steps:

1. Determine the exposed range of the sliding block on the dam foundation surface.

2. Calculate the resultant of all actions of the arch dam within the range.

C.4.2 The self-weight of a sliding block shall be the product of the sliding block volume and the rock unit weight. The sliding block volume shall exclude the volume of excavated rocks.

C.4.3 The uplift pressure on a sliding block shall be calculated in accordance with Appendix A of this code.

Appendix D Single Safety Factor Design for Sliding Stability of Abutment Under Non-seismic Loading

D.0.1 The sliding stability calculation of abutment in non-seismic conditions shall consider the dam thrust, self-weight of sliding block, uplift pressure, etc. The loads on the sliding block shall be calculated using the method in Section C.4 of this code. The design load combination shall include the fundamental combination and the special combination. The fundamental combination shall be determined according to the persistent and transient situations in Table 5.2.2 of this code, and the special combination shall be determined according to the non-seismic accidental situations in Table 5.2.2 of this code.

D.0.2 When the stability of the abutment in non-seismic conditions is analyzed by the rigid limit equilibrium method, Formula (D.0.2-1) shall be used for arch dams of Grades 1 and 2, as well as the high dams, and Formula (D.0.2-1) or (D.0.2-2) shall be used for other arch dams.

$$K_1 = \frac{\sum (f'N + c'A)}{\sum T} \qquad (\text{D.0.2-1})$$

$$K_2 = \frac{\sum fN}{\sum T} \qquad (\text{D.0.2-2})$$

where

K_1 is the safety factor for sliding stability calculated based on the peak shear strength;

K_2 is the safety factor for sliding stability calculated based on the residual shear strength;

f' is the shear-friction coefficient, which shall consider the average value of the peak strength of material;

f is the friction coefficient, which considers proportion limit strength for brittle failure materials, yield strength for plastic or brittle-plastic failure materials, and residual strength for shear broken materials;

c' is the cohesion (MPa), which shall consider the average value of the peak strength of material;

N is the normal force perpendicular to the sliding plane (10^3 kN);

A is the sliding plane area (m²);

T is the sliding force in the sliding direction (10^3 kN).

D.0.3 The single safety factor of the sliding stability of the abutment calculated by the rigid limit equilibrium method in non-seismic conditions shall not be less than the values specified in Table D.0.3.

Table D.0.3 Single safety factor of sliding stability of abutment in non-seismic conditions

Calculation formula	Load combination		Structure grade			
			1	2	3	4, 5
(D.0.2-1)	Fundamental combination	Persistent condition	3.50	3.25	3.25	3.00
		Transient condition	3.35	3.10	3.10	2.85
	Special combination	Accidental condition	3.00	2.75	2.75	2.50
(D.0.2-2)	Fundamental combination	Persistent condition	–	–	1.30	1.25
		Transient condition	–	–	1.25	1.20
	Special combination	Accidental condition	–	–	1.15	1.10

Appendix E　Hydraulic Formulae

E.1　Weir Surface Curve of Overflow Spillway

E.1.1 The surface curve of an open overflow weir shall meet the following requirements:

1. An exponential curve should be adopted for the surface curve downstream of the origin of an open overflow weir, which is calculated by the following formulae:

$$x^n = kH_s^{n-1}y \qquad (E.1.1\text{-}1)$$

$$H_s = (0.75 \sim 0.95)H_{zmax} \qquad (E.1.1\text{-}2)$$

where

- x, y　are the coordinates with the origin at the apex of weir crest (m). x is positive in the downstream direction, whereas y is positive in the downward direction;
- k, n　are the weir surface curve parameters, taken from Table E.1.1-1;
- H_s　is the design head (m), which may be determined according to the allowable negative pressure on the weir surface. The possible maximum negative pressure on the weir crest may be obtained by consulting Table E.1.1-2;
- H_{zmax}　is the elevation difference between the highest pool level and the weir crest (m).

Table E.1.1-1　Surface curve parameters of overflow spillway

Vertical upstream surface	k	n
	2.00	1.85

Table E.1.1-2　Possible maximum negative pressure on weir crest of overflow spillway

H_s / H_{zmax}	0.750	0.775	0.800	0.825	0.850	0.875	0.900	0.950	1.000
Maximum negative pressure (9.81 kPa)	0.500 H_s	0.450 H_s	0.400 H_s	0.350 H_s	0.300 H_s	0.250 H_s	0.200 H_s	0.100 H_s	0.000 H_s

2. When the upstream dam surface is vertical, the elliptic curve and tri-arc curve may be adopted for the surface curve upstream of the origin of

the weir.

1) The elliptic curve upstream of the origin of the weir (Figure E.1.1-1) can be calculated by the following formulae:

$$\frac{x^2}{(aH_s)^2} + \frac{(bH_s - y)^2}{(bH_s)^2} = 1 \qquad (E.1.1-3)$$

$$a/b = 0.87 + 3a \qquad (E.1.1-4)$$

where

    a, b are the coefficients, where a is taken as 0.28 to 0.30;

    aH_s, bH_s are the major semi-axis A and minor semi-axis B of the ellipse, respectively (m).

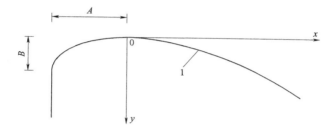

Key

A major semi-axis of ellipse

B minor semi-axis of ellipse

1 exponential curve downstream of the origin

Figure E.1.1-1 Elliptic curve upstream of the origin of an open overflow weir

2) The parameters of the tri-arc curve upstream of the origin of an open overflow weir (Figure E.1.1-2) shall be in accordance with Table E.1.1-3.

Table E.1.1-3 Parameters of tri-arc curve upstream of the origin of an open overflow weir

R_1	R_2	R_3	L_1	L_2	L_3
$0.5H_s$	$0.2H_s$	$0.04H_s$	$0.175H_s$	$0.276H_s$	$0.282H_s$

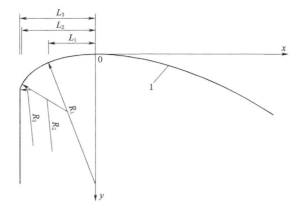

Key

R_1, R_2, R_3 radii of three arcs
L_1, L_2, L_3 horizontal distances from the upstream end points of three arcs to the origin, respectively
1 exponential curve downstream of the origin

**Figure E.1.1-2   Tri-arc curve upstream of the origin
of an open overflow weir**

    3   The height of the overhang of an open overflow weir (Figure E.1.1-3) shall satisfy the following formula:

$$d > \frac{H_{zmax}}{2} \quad\quad\quad (E.1.1\text{-}5)$$

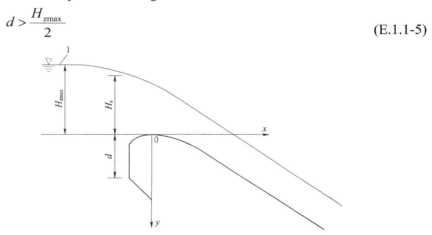

Key

1 highest pool level
d height of the overhang of an open overflow weir

Figure E.1.1-3   Overhang of an open overflow weir

E.1.2   The design of surface curve of overflow spillway, with breast wall or whose flow is orifice discharge when the gate is fully open, shall meet the following requirements:

    1   A single-circle, multiple-circle or elliptic curve may be adopted for the weir surface curve upstream of the origin, which may be selected considering the bottom shape of breast wall.

2   Weir surface curve downstream from the origin for orifice discharge (Figure E.1.2) should be determined by test when the ratio of the maximum working head to the height of the orifice is 1.2 to 1.5, and can be calculated by the following formulae when the ratio is greater than 1.5:

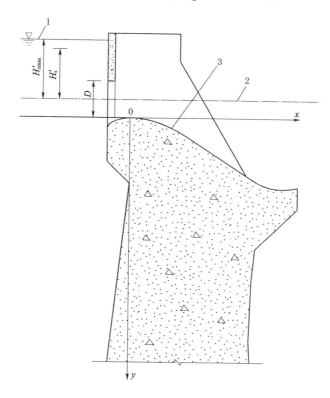

Key

1   highest pool level
2   centerline of orifice
3   weir surface curve downstream from the origin
D   orifice height

Figure E.1.2   Weir surface curve downstream from the origin for orifice discharge

$$y = \frac{x^2}{4\varphi^2 H'_s} \tag{E.1.2-1}$$

$$H'_s = (0.75 \sim 0.95) H'_{zmax} \tag{E.1.2-2}$$

where

φ is the flow velocity coefficient at the contraction section of the orifice, which may be taken as 0.96, or 0.95 for the orifice with a bulkhead gate slot;

H'_s is the design head (m);

H'_{zmax} is the maximum working head of the orifice (m), which is the maximum working head on the centerline of the orifice.

E.2 Discharge Capacity

E.2.1 The discharge capacity of an open overflow weir shall be calculated by the following formula:

$$Q = Cm\varepsilon\sigma_s B\sqrt{2g}H_z^{3/2} \qquad (E.2.1)$$

where

Q is the discharge (m³/s);

C is the correction coefficient for the influence of upstream surface slope, taken as 1.0 for the vertical upstream surface;

m is the discharge coefficient, taken from Table E.2.1;

ε is the lateral contraction coefficient determined according to the pier thickness and pier nose shape, which may be taken as 0.90 to 0.95;

σ_s is the submergence coefficient determined according to the submergence degree of the discharge, taken as 1.0 for non-submerged flow;

B is the net width of open overflow weir (m);

g is the gravity acceleration (m/s²);

H_z is the working head on the weir crest (m).

Table E.2.1 Discharge coefficient

H_z/H_s	P/H_s				
	0.20	0.40	0.60	1.00	≥1.33
0.4	0.425	0.430	0.431	0.433	0.436
0.5	0.438	0.442	0.445	0.448	0.451
0.6	0.450	0.455	0.458	0.460	0.464
0.7	0.458	0.463	0.468	0.472	0.476
0.8	0.467	0.474	0.477	0.482	0.486
0.9	0.473	0.480	0.485	0.491	0.494
1.0	0.479	0.486	0.491	0.496	0.501
1.1	0.482	0.491	0.496	0.502	0.507
1.2	0.485	0.495	0.499	0.506	0.510
1.3	0.496	0.498	0.500	0.508	0.513

NOTE P is the elevation difference between the weir crest and reservoir bottom (m).

E.2.2 The discharge coefficient for the long pressure outlet in dam shall be calculated by the following formula:

$$\mu = \frac{1}{\sqrt{1 + \sum \xi_{li}\left(\frac{A_k}{A_i}\right)^2 + \sum \xi_{Mi}\left(\frac{A_k}{A_i}\right)^2}} \tag{E.2.2}$$

where

- μ is the discharge coefficient for long pressure outlet in dam;
- ξ_{li} is the local head loss coefficient, mainly including head loss coefficients of inlet, transition section, gate slot and bend;
- ξ_{Mi} is the frictional head loss coefficient;
- A_k is the control cross-section area at the exit (m²);
- A_i is the cross-section area corresponding to ξ_{li} and ξ_{Mi} (m²).

E.2.3 The discharge capacity of outlet shall be calculated by the following formula:

$$Q = \mu A_k \sqrt{2gH_z} \tag{E.2.3}$$

where

- μ is the discharge coefficient for outlet or conduit, which is taken as 0.83 to 0.93 for short pressure outlets, and shall be calculated for long pressure outlets in accordance with Article E.2.2 of this code;
- H_z is the working head of outlet (m), taken as the working head at the centerline of outlet when the free discharge boundary is subjected to atmospheric pressure, the working head at the top edge of outlet when the free discharge has a bottom boundary, or the difference between upstream and downstream water levels when the discharge is submerged.

E.3 Hydraulic Factors for Ski-Jump Energy Dissipation

E.3.1 The hydraulic factors of ski-jump energy dissipation (Figure E.3.1) shall include the maximum scour pit depth and the trajectory distance.

E.3.2 The maximum scour pit depth and trajectory distance can be calculated by the following formulae:

$$v_1 = 1.1\varphi\sqrt{2g(H_0 - h_1 \cos\theta)} \tag{E.3.2-1}$$

$$L = \frac{1}{g}[v_1^2 \sin\theta \cos\theta + v_1 \cos\theta\sqrt{v_1^2 \sin^2\theta + 2g(h_1 \cos\theta + h_2)}] \tag{E.3.2-2}$$

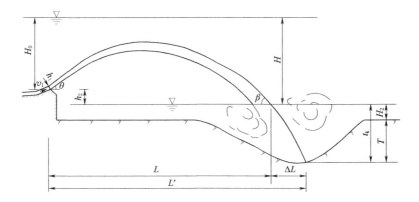

Key

L　trajectory distance, i.e., horizontal distance from the flip bucket terminal to the outer edge of the jet nappe intersecting the downstream water surface

L'　horizontal distance between the flip bucket terminal and the lowest point of the scour pit

ΔL　horizontal distance from the outer edge of the jet nappe to the lowest point of the scour pit

H_0　elevation difference between pool level and bucket lip

H　difference between upstream and downstream water levels

H_2　downstream water depth

θ　bucket flip angle

v_1　surface velocity of flow at the bucket lip

h_1　average water depth normal to the bucket lip

h_2　elevation difference between the bucket lip and the downstream water surface

β　angle between the outer edge of jet nappe and the downstream water surface

t_k　depth from downstream water surface to the lowest point of the scour pit

T　maximum scour pit depth

Figure E.3.1　Hydraulic factors for ski-jump energy dissipation

$$\tan\beta = \sqrt{\tan^2\theta + \frac{2g(h_2 + h_1\cos\theta)}{v_1^2 \cos\theta}} \tag{E.3.2-3}$$

$$t_k = kq^{0.5}H^{0.25} \tag{E.3.2-4}$$

$$T = t_k - H_2 \tag{E.3.2-5}$$

$$\Delta L = t_k \tan^{-1}\beta \tag{E.3.2-6}$$

$$L' = L + \Delta L \tag{E.3.2-7}$$

where

　　v_1　is the surface velocity of flow at the bucket lip (m/s);

　　φ　is the weir surface flow velocity coefficient, which may be taken as 0.95 for initial estimate;

H_0 is the elevation difference between the pool level and bucket lip (m);

h_1 is the average water depth normal to the bucket lip (m);

θ is the bucket flip angle (°);

L is the trajectory distance (m), which is the horizontal distance from the flip bucket terminal to the outer edge of the jet nappe intersecting the downstream water surface;

h_2 is the elevation difference between the bucket lip and the downstream water level (m);

β is the jet trajectory impact angle (°), which is the included angle between the jet nappe outer edge and the downstream water surface;

t_k is the water depth from downstream water surface to the lowest point of the scour pit (m);

k is the bedrock scour coefficient, taken from Table E.3.2 according to the characteristics of bedrock;

q is the discharge per unit width at the flip bucket [m³/(s·m)];

H is the difference between upstream and downstream water levels (m);

T is the maximum scour pit depth (m), which is the depth from riverbed surface to pit bottom;

H_2 is the downstream water depth (m);

L' is the horizontal distance between the flip bucket terminal and the lowest point of the scour pit.

Table E.3.2 Bedrock scour coefficient

Scouring feature		Difficult	Prone to scour	Relatively easy	Easy
Joints and fissures	**Space** (cm)	> 150	50 to 150	20 to 50	< 20
	Development degree	Undeveloped, 1 to 2 joint (fissure) sets, regular	Relatively developed, 2 to 3 joint (fissure) sets, X-shaped, relatively regular	Developed, more than 3 joint (fissure) sets, in "X" or "*" shape, irregular	Well developed, more than 3 joint (fissure) sets, messy, cut into crushed rock

Table E.3.2 *(continued)*

Scouring feature		Difficult	Prone to scour	Relatively easy	Easy
Structural characteristics of bedrocks	Intactness	Huge block	Large block	Blocky and crushed rock	Crushed rock
	Structure type	Integral	Block	Mosaic	Cataclastic
	Fissure property	Original or structural cracks, mostly closed, with short persistence	Mainly composed of structural cracks, mostly closed, partly open, with less filling and well cemented	Mainly composed of structural or weathered cracks, mostly slightly open and partly open, partly filled with clay and poorly cemented	Mainly composed of structural or weathered cracks, with fissures slightly open or open, partly filled with clay and poorly cemented
k	Range	0.6 to 0.9	0.9 to 1.2	1.2 to 1.6	1.6 to 2.0
	Mean	0.8	1.1	1.4	1.8

NOTE Applicable to a jet trajectory impact angle of 30° to 70°.

E.4 Free-Fall Jet Flow Energy Dissipation Factors

E.4.1 Free-fall jet flow energy dissipation factors (Figure E.4.1) shall include the trajectory distance, water depth upstream of the impact point, impact flow velocity on the apron, hydrodynamic pressure, and maximum scour depth.

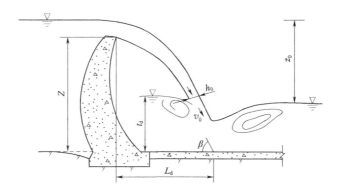

Key

L_d trajectory distance

Z elevation difference from bucket to riverbed

Z_0 elevation difference between upstream and downstream water levels

t_d water depth upstream of the impact point

h_0 thickness of jet nappe at the water surface

v_0 average flow velocity of jet when it falls down to water surface

$β$ impact angle

Figure E.4.1 Free-fall jet flow energy dissipation factors

E.4.2 The trajectory distance for free-fall energy dissipation can be calculated by the following formula:

$$L_d = 2.3 q^{0.54} Z^{0.19} \tag{E.4.2}$$

where

L_d is the trajectory distance (m);

q is the discharge per unit width at bucket [m³/(s·m)];

Z is the elevation difference between bucket and riverbed (m).

E.4.3 The water depth upstream of the impact point can be calculated by the following formula:

$$t_d = 0.6 q^{0.44} Z^{0.34} \tag{E.4.3}$$

where

t_d is the water depth upstream of the impact point.

E.4.4 The impact flow velocity on the apron can be calculated by the following methods:

1 When there is difference between the upstream and downstream water levels at the impact point, the impact flow velocity on the apron can be calculated by the following formula:

$$v_1 = 4.88 q^{0.15} Z^{0.275} \tag{E.4.4-1}$$

where

v_1 is the impact flow velocity on the apron (m/s).

2 When the jet is distributed almost all over the river surface and there is no noticeable difference between the upstream and downstream water levels at the impact point, the impact flow velocity on the apron can be calculated by the following formulae:

$$v_0 = \varphi \sqrt{2gZ_0} \tag{E.4.4-2}$$

$$\beta = \cos^{-1}\left(\frac{2v_1}{v_0} - 1\right) \tag{E.4.4-3}$$

$$h_0 = \frac{q}{v_0} \tag{E.4.4-4}$$

$$v_1 = \frac{2.5 v_0}{\sqrt{\dfrac{t_d}{h_0 \sin \beta}}} \tag{E.4.4-5}$$

where

- v_0 is the average flow velocity of jet nappe at the water surface (m/s);
- φ is the flow velocity coefficient;
- Z_0 is the difference between the upstream and downstream water levels (m);
- β is the impact angle;
- h_0 is the thickness of jet nappe at the water surface.

E.4.5 The hydrodynamic pressure on the apron can be calculated by the following formula:

$$P_\mathrm{d} = \frac{\gamma(v_1 \sin\beta)^2}{2g} \tag{E.4.5}$$

where

- P_d is the hydrodynamic pressure (kN/m^2);
- γ is the unit weight of water (kN/m^3).

E.4.6 In the case of no apron, the maximum scour pit depth can be calculated in accordance with Section E.3 of this code, where the larger value of scour coefficient should be taken.

E.5 Flow Cavitation Number

E.5.1 Flow cavitation number can be calculated by the following formulae:

$$h_\mathrm{d} = 10.33 - \nabla/900 \tag{E.5.1-1}$$

$$\sigma = \frac{h_0 + h_\mathrm{d} - h_\mathrm{v}}{v_0^2/2g} \tag{E.5.1-2}$$

where

- σ is the flow cavitation number;
- h_0 is the time-averaged hydrodynamic pressure head at the calculation section (m). When the flow velocity is greater than 30 m/s, the effect of fluctuating pressure shall be considered;
- h_d is the atmospheric pressure head at the calculation section (m);
- h_v is the water vaporization pressure head (m), which may be determined based on the water temperature according to Table E.5.1;
- v_0 is the average flow velocity at the calculation section (m/s);

∇ is the height above sea level (m).

Table E.5.1 Water vaporization pressure head

Water temperature (°C)	0	5	10	15	20	25	30	40
h_v (m)	0.06	0.09	0.13	0.17	0.24	0.32	0.43	0.75

Appendix F Classification and Selection of Monitoring Items

Table F Classification and selection of monitoring items

No.	Monitoring category	Monitoring item	Dam classification by height			
			Extra-high dam	High dam	Medium dam	Low dam
I	Patrol inspection	Structures, slopes and near-dam banks in project area	●	●	●	●
II	Deformation	1 Dam displacement	●	●	●	●
		2 Dam chord length	●	●	○	○
		3 Abutment displacement	●	●	●	●
		4 Valley deformation at abutments and its upstream and downstream range. Deep deformation of rock mass	●	●	○	○
		5 Inclination	●	●	○	○
		6 Deformation of joints	●	●	●	●
		7 Deformation of cracks	●	●	●	●
		8 Foundation displacement	●	●	●	●
		9 Displacement of near-dam banks	●	●	○	○
		10 Valley deformation of the near-dam banks. Deep deformation of rock mass	●	●	○	○
III	Seepage	1 Seepage quantity	●	●	●	●
		2 Foundation seepage pressure	●	●	●	●
		3 Dam seepage pressure	●	○	○	○
		4 Bypass seepage (including groundwater level)	●	●	●	●
		5 Water quality analysis	●	○	○	○

Table F *(continued)*

No.	Monitoring category	Monitoring item	Dam classification by height			
			Extra-high dam	High dam	Medium dam	Low dam
IV	Stress, strain and temperature	1 Dam stress and strain	●	●	○	○
		2 Foundation stress and strain	●	●	○	○
		3 Concrete temperature	●	●	●	●
		4 Foundation temperature	●	●	●	○
		5 Stress of pre-stressed anchor cable in gate pier	●	●	●	●
		6 Stress of rebar in gate pier	●	●	○	○
		7 Stress of outlet rebar	●	○	○	○
V	Environmental quantities	1 Upstream and downstream water levels	●	●	●	●
		2 Air temperature	●	●	●	●
		3 Precipitation	●	●	●	●
		4 Reservoir water temperature	●	●	○	○
		5 Sedimentation in front of the dam	●	●	○	○
		6 Downstream scour	●	○	○	○
		7 Freezing	○	○	○	○
		8 Atmospheric pressure	●	○	○	○

NOTES:

1 "●" denotes mandatory, "○" denotes optional, depending on the project needs.
2 The deformation monitoring of dam cracks shall be arranged according to the crack conditions at the construction stage.
3 Upstream and downstream water level monitoring may be performed by the hydrological telemetry and forecasting system.

Explanation of Wording in This Code

1. Words used for different degrees of strictness are explained as follows in order to mark the differences in executing the requirements in this code:

    1) Words denoting a very strict or mandatory requirement:

        "Must" is used for affirmation; "must not" for negation.

    2) Words denoting a strict requirement under normal condition:

        "Shall" is used for affirmation; "shall not" for negation.

    3) Words denoting a permission of a slight choice or in an indication of the most suitable choice when conditions permit:

        "Should" is used for affirmation; "should not" negation.

    4) "May" is used to express the option available, sometimes with the conditional permit.

2. "Shall meet the requirements of …" or "shall comply with…" or is used in this code to indicate that it is necessary to comply with the requirements stipulated in other relative standards and codes.

List of Quoted Standards

GB 50201,	*Standard for Flood Control*
GB 50872,	*Code for Fire Protection Design of Hydropower Projects*
GB/T 51394,	*Standard for Load on Hydraulic Structures*
GB 748,	*Sulfate Resistance Portland Cement*
GB 4053.1,	*Safety Requirements for Fixed Steel Ladders and Platform - Part 1: Steel Vertical Ladders*
NB/T 10335,	*Code for Design of Roller-Compacted Concrete Arch Dams*
NB/T 10512,	*Code for Slope Design of Hydropower Projects*
NB/T 10857,	*Design Code for Reasonable Service Life and Durability of Hydropower Projects*
NB/T 10867,	*Code for Design of Spillways*
NB/T 35024,	*Design Code for Hydraulic Structures Against Ice and Freezing Action*
NB/T 35026,	*Design Code for Concrete Gravity Dams*
NB 35047,	*Code for Seismic Design of Hydraulic Structures of Hydropower Project*
NB 35055,	*Design Code for Steel Gate in Hydropower Projects*
NB/T 35092,	*Design Code for Temperature Control of Concrete Dam*
DL 5180,	*Classification & Design Safety Standard of Hydropower Projects*
DL/T 1558,	*Code for Operation and Maintenance of Dam Safety Monitoring System*
DL/T 5057,	*Design Specification for Hydraulic Concrete Structures*
DL/T 5100,	*Technical Code for Chemical Admixtures for Hydraulic Concrete*
DL/T 5144,	*Specifications for Hydraulic Concrete Construction*
DL/T 5148,	*Technical Specification for Cement Grouting Construction of Hydraulic Structures*
DL/T 5150,	*Test Code for Hydraulic Concrete*
DL/T 5178,	*Technical Specification for Concrete Dam Safety Monitoring*
DL/T 5207,	*Technical Specification for Abrasion and Cavitation Resistance*

	of Concrete in Hydraulic Structures
DL/T 5209,	*Specification of Information Compilation for Concrete Dam Safety Monitoring*
DL/T 5215,	*Specification for Waterstop for Hydraulic Structures*
DL/T 5313,	*Guide for Safety Assessment of Large Dams for Hydropower Station in Operation*
DL/T 5330,	*Code for Mix Design of Hydraulic Concrete*